# Anatomy & Physiology

3rd Edition

## by Erin Odya and Maggie Norris

for
dummies®
A Wiley Brand

## Anatomy & Physiology For Dummies®, 3rd Edition

Published by: **John Wiley & Sons, Inc.,** 111 River Street, Hoboken, NJ 07030-5774, www.wiley.com

Copyright © 2017 by John Wiley & Sons, Inc., Hoboken, New Jersey

Published simultaneously in Canada

No part of this publication may be reproduced, stored in a retrieval system or transmitted in any form or by any means, electronic, mechanical, photocopying, recording, scanning or otherwise, except as permitted under Sections 107 or 108 of the 1976 United States Copyright Act, without the prior written permission of the Publisher. Requests to the Publisher for permission should be addressed to the Permissions Department, John Wiley & Sons, Inc., 111 River Street, Hoboken, NJ 07030, (201) 748-6011, fax (201) 748-6008, or online at http://www.wiley.com/go/permissions.

Trademarks: Wiley, For Dummies, the Dummies Man logo, Dummies.com, Making Everything Easier, and related trade dress are trademarks or registered trademarks of John Wiley & Sons, Inc., and may not be used without written permission. All other trademarks are the property of their respective owners. John Wiley & Sons, Inc., is not associated with any product or vendor mentioned in this book.

For general information on our other products and services, please contact our Customer Care Department within the U.S. at 877-762-2974, outside the U.S. at 317-572-3993, or fax 317-572-4002. For technical support, please visit https://hub.wiley.com/community/support/dummies.

Wiley publishes in a variety of print and electronic formats and by print-on-demand. Some material included with standard print versions of this book may not be included in e-books or in print-on-demand. If this book refers to media such as a CD or DVD that is not included in the version you purchased, you may download this material at http://booksupport.wiley.com. For more information about Wiley products, visit www.wiley.com.

Library of Congress Control Number: 2017932045

ISBN 978-1-119-34523-7 (pbk); ISBN 978-1-119-34530-5 (ebk); ISBN 978-1-119-34536-7 (ebk)

Manufactured in the United States of America

10  9  8  7  6  5  4  3  2

# Contents at a Glance

# Table of Contents

# Introduction

Congratulations on your decision to study human anatomy and physiology. The knowledge you gain from your study is of value in many aspects of your life.

Begin with the most obvious: the social value of this knowledge. Human anatomy and physiology is always a suitable topic of discussion in social situations because it allows people to talk about their favorite subject (themselves) in a not-too-personal way. Thus, some particularly interesting detail of anatomy and physiology is an ideal conversation opener with attractive strangers or horrifying shirt-tail relatives. (First, though, be completely clear in your mind about the boundary between scientific anatomy and physiology on the one hand and personal clinical details on the other.) Choose the specific topic carefully to be sure of having your intended effect. For example, telling a young boy that he has the same density of hair follicles on his body as a chimp does will probably please him. Telling his teenage sister the same thing may alienate her. Use this power carefully!

A little background in anatomy and physiology should be considered a valuable part of anyone's education. Health and medical matters are part of world events and people's daily lives. Basic knowledge of anatomy and physiology gets you started when trying to make sense of the news about epidemics, novel drugs and medical devices, and purported environmental hazards, to name just a few examples. Anatomy and physiology prepare you to be a more well-rounded, knowledgeable person and will help you be a better parent, spouse, care-giver, neighbor, friend, or colleague.

Knowledge of anatomy and physiology may also benefit your own health. Sometimes, comprehension of a particular fact or concept can help drive a good decision about long-term health matters, like the demonstrated benefits of exercise, or it may help you take appropriate action in the context of a specific medical problem, like an infection, an infarction, a cut, or a muscle strain. You may understand your doctors' instructions better during a course of treatment, which may give you a better medical outcome.

# About This Book

This book guides you on a quick walk-through of human anatomy and physiology. It doesn't have the same degree of technical detail as a textbook. It contains relatively little in the way of lists of important anatomical structures, for instance.

We expect that most readers are using this book as a complementary resource for course work in anatomy and physiology at the high-school, college, or career-training level. Most of the information overlaps with the information available in your other resources. However, sometimes a slightly different presentation of a fact or of the relationship between facts can lead to a small "aha!" Some technical details in your more comprehensive resources may become easier to master after that. Consider reading the relevant chapter prior to class. That way, when your instructor covers the content, it'll be more likely to stick!

The goals of this book are to be informal but not unscientific; brief but not sketchy; and information-rich but accessible to readers at many levels. We've tried to present a light but serious survey of human anatomy and physiology that you can enjoy for the sake of the information it imparts and that will help you perform well on your tests. As always, the reader is the judge of its success.

You won't find clinical information in this book. Chapters 4 through 15 have a pathophysiology section that uses disorders and disease states to explore the details of some physiological processes, but this book contains nothing related to patient care or self-care. It's also not a health and wellness manual or any kind of lifestyle book.

Within this book, you may note that some web addresses break across two lines of text. If you're reading this book in print and want to visit one of these web pages, simply key in the web address exactly as it's noted in the text, pretending as though the line break doesn't exist. If you're reading this as an e-book, you've got it easy — just click the web address to be taken directly to the web page.

# Foolish Assumptions

When we wrote this book, we tried to keep you in mind. We're guessing that you fall into one of these categories:

>> **Formal student:** You're a high-school or college student enrolled in a basic anatomy and physiology course for credit, or a student in a career-training program for a certification or credential. You need to pass an exam or

otherwise demonstrate understanding and retention of data, terminology, and concepts in human anatomy and physiology.

>> **Informal student:** You're not enrolled in a credit course, but gaining some background in human anatomy and physiology is important to you for personal or professional reasons.

>> **Casual reader:** Here you are with a book on your hands and a little time to spend reading it. And it's all about you!

# Icons Used in This Book

The little round pictures that you see in the margins throughout this book are icons that alert you to several different kinds of information.

TIP

The Tip icon lets you know what you can do to improve your understanding of an anatomical structure.

REMEMBER

The Remember icon serves to jog your memory. Sometimes, the text is information that we think you should permanently store in your anatomy and physiology file. Other times, the info here makes a connection between what you're reading and related information elsewhere in the book.

TECHNICAL
STUFF

The Technical Stuff icon flags extra information that takes your understanding of anatomy or physiology to a slightly deeper level, but the text isn't essential for understanding the organ system under discussion.

# Beyond the Book

In addition to the material in the print or e-book you're reading right now, this product also comes with some access-anywhere goodies on the web. Check out the free Cheat Sheet for more on everything from anatomical terms to the anatomical planes of the body and more. To get this Cheat Sheet, simply go to www.dummies.com and type **Anatomy & Physiology For Dummies Cheat Sheet** in the Search box.

# Where to Go from Here

If you're a formal student (that is, one who's enrolled or planning to enroll in a formal course in human anatomy and physiology), you may get the most benefit by becoming familiar with this book a week or two before your course begins. Flip to the color plates in the center of the book to get started. The illustrations, charming as well as scientific, are arranged to follow the flow of the text, and the callouts indicate important technical terminology.

Then peruse the book as you would any science book; look at the table of contents and the index. Read the Introduction. (See, you've started already!) Then start reading chapters. Look at the figures, especially the color plates, as you read. You'll probably be able to get through the entire book in just a couple of sittings. Then go back and reread chapters you found particularly interesting, relevant, or puzzling. Study the illustrations carefully. The line drawings as well as the color plates are keyed closely to the text and often clarify important facts. Pay attention to technical terminology; your instructors will use it and expect you to use it, too.

If you're a casual reader (you're not enrolled in a formal course in anatomy and physiology and have little or no background in biology), the following approach may work well. Take some time with the color plates at the center of the book. They give you a good feel for the flow of information (and a good feeling about the human body). Then read the book straight through, beginning to end. Look at the figures, especially the color plates, as you read. After you've been through it all quickly once, go back and reread chapters you found particularly interesting, relevant, or puzzling. Make a habit of studying the illustrations while reading the related text. Don't sweat too much over terminology; for your purposes, saying "of my lungs" communicates as well as "pulmonary." (If you also enjoy word games, though, you can get started on a whole new vocabulary.) Keep the book handy for future reference the next time you wonder what the heck they're talking about in a TV drug ad. The color plates alone make it worth space on your bookshelf.

# 1
# Locating Physiology on the Web of Knowledge

Get acquainted with the basics of anatomy and physiology.

Find out about metabolism — all the chemical reactions that keep you alive.

Learn how we keep everything in check — maintaining balance in our bodies.

Brush up on biochemistry.

Find the fundamentals of cell biology.

See how cells organize into tissues.

# Chapter **1**

# Anatomy and Physiology: The Big Picture

Human *anatomy* is the study of the human body's structures — all the parts that make up the physical body itself. *Physiology* is the study of how the human body works; how all the anatomical parts function together to keep an individual alive. Anatomy and physiology are bound together. As such, this book abandons the old technique of learning all the anatomy and then the physiology as though the two were independent. Here, we examine each body system, identify the structures within that system, and then discuss their functions.

## Scientifically Speaking

Human anatomy and physiology are closely related to *biology*, which is the study of living things and their relationship with the rest of the universe, including all other living things. If you've studied biology, you understand the basics of how organisms operate. Anatomy and physiology narrow the science of biology by looking at the specifics of one species: *Homo sapiens*.

REMEMBER

Anatomy is form; physiology is function. You can't talk about one without talking about the other.

## THE ANATOMY AND PHYSIOLOGY OF EVERYTHING ELSE

Scientifically speaking, human biology isn't more or less complex, specialized, or cosmically significant than the biology of any other species, and all are interdependent. Every species of animal, plant, and fungus on the planet has both anatomy and physiology. So does each species of *protist* (one-celled creatures, like amoebae) and bacteria. At the cellular level (see Chapter 3), all these groups are astoundingly similar. At the levels of tissues, organs, and organ systems, plants are very different from animals, and both plants and animals are equally dissimilar to fungi.

Each of these major groups, called a *kingdom,* has its own characteristic anatomy and physiology. It's evident at a glance to everyone at the beach that a starfish and a human are both animals, while the seaweed in the tide pool and the cedar tree on the shoreline are both plants. Obvious details of anatomy (the presence or absence of bright green tissue) and physiology (the presence or absence of movement) tell that story. The different forms within each kingdom have obvious differences as well: The cedar must stand on the shore, but the seaweed would die there. The starfish can move from one place to another within a limited range, while humans can (theoretically) go anywhere on the planet and survive there for at least a while. Scientists use these differences to classify organisms into smaller and smaller groups within the kingdom, until each organism is classified into its own special group.

Not that human anatomy and physiology aren't special. Humans' bipedal posture and style of locomotion are very special . There's nothing like a human hand anywhere except at the end of a human arm. Perhaps most special of all is the anatomy and physiology that allows (or maybe compels) humans to engage in science: our highly developed brain and nervous system. It's entirely within the norms of evolutionary theory that people would be most interested in their own species, so more humans find human anatomy and physiology more interesting than the anatomy and physiology of the tree. From here on, we're restricting our discussion to the anatomy and physiology of our own species.

## How anatomy and physiology fit into science

Biologists base their work on the assumption that every structure and process, no matter how tiny in scope, must somehow contribute to the survival of the individual. So each process — and the chemistry and physics that drive it — must help keep the individual alive and meeting the relentless challenges of a continually changing environment. Although anatomy and physiology combined are classified as a subsection of biology, it's truly an interdisciplinary science.

Human *pathophysiology* is the study of "human anatomy and physiology gone wrong." (The prefix *path-* is Greek for "suffering.") It's the interface of human biology and medical science. *Clinical medicine* is the application of medical science to alleviate an anatomical or physiological problem in an individual human.

Pathophysiology and clinical medicine aren't the subject of this book, but we discuss applications of them when they're particularly relevant to the physiology. You're probably using this book to supplement instructional material in career training for a clinical environment, so the information throughout the book is slightly slanted in that direction. We chose the conditions that we briefly examine to demonstrate some characteristic of the system, especially its interaction with other systems, but we don't discuss diagnosis or treatment.

## TAXONOMY OF *HOMO SAPIENS*

*Taxonomy* is the science that seeks to classify and organize living things, expressed as a series of mutually exclusive categories. The highest (most inclusive) category is domain, of which there are three: Archea, Eubacteria, Eukaryota. Each of these domains is split into kingdoms, which are further divided until each individual organism is its own unique species. Outside of bacteria, all living things fall under the Eukaryota domain; the kingdoms are: Protista, Fungi, Plantae, and Animalia. Within each kingdom, the system classifies each organism into the hierarchical subgroups (and sometimes sub-subgroups) of phylum, class, order, family, genus, and species. Here's the breakdown of humankind:

**Kingdom Animalia:** All animals.

**Phylum Chordata:** Animals that have a number of structures in common, particularly the *notochord,* a rodlike structure that forms the body's supporting axis.

**Subphylum Vertebrata:** Animals with backbones.

**Superclass Tetrapoda:** Four-footed vertebrates.

**Class Mammalia:** Tetrapods with hair. Other classes of the vertebrata are Pisces (fish), Amphibia (frogs), Aves (birds), and Reptilia (scaly things).

**Order Primates:** Mammals with more highly developed brains, flexible hips and shoulders, and prehensile hands and feet (able to grasp).

**Superfamily Hominoidea:** Apes (chimpanzees, gorillas, orangutans, humans).

*(continued)*

*(continued)*

**Family Hominidae:** Great apes, including humans.

**Genus Homo:** The human species is the only surviving species of our genus, though this genus included several species in the evolutionary past.

**Species Sapiens:** All species are given a two-part Latin name, in which the genus name comes first and a species epithet comes second. The biologists who name species sometimes try to use a descriptor in the epithet. For humans, they could have chosen "bipedal" or "talking" or "hairless," but they chose "thinker."

**Variety Sapiens:** Some species get a "varietal" name, usually indicating a difference that's obvious but not necessarily important from an evolutionary point of view. The human species has one other variety, *Homo sapiens neanderthalensis,* which has been extinct for tens of thousands of years. All humans living since then are of one species variety, *Homo sapiens sapiens.* In the evolutionary classification of humans, there's no biologically valid category below species variety.

# Anatomy, gross and otherwise

Some biologists specialize in the anatomy and physiology of animals at various hierarchical levels (horses, fish, frogs) or particular organs (mammalian circulatory systems, olfaction in fish, insect hormones). Some focus solely on humans, others concentrate on other species, and still others examine the areas of overlap between humans and other animal species. These various areas of study contribute to our knowledge of biology in general and have important applications in clinical medicine. The work of anatomists contributes to medical advances, such as improved surgical techniques and the development of bioengineered prostheses.

Throughout this book, you encounter some information from each major subset of anatomy, including

>> **Gross anatomy:** The study of the large parts of an animal body — any animal body — that can be seen with the unaided eye. That's the aspect of anatomy we concentrate on in this book.

>> **Histologic anatomy:** The study of different tissue types and the cells that comprise them. Histologic anatomists use a variety of microscopes to study the cells and tissues that make up the body.

>> **Developmental anatomy:** The study of the life cycle of the individual, from fertilized egg through adulthood, senescence (aging), and death. Body parts

change throughout the life span. For information about human developmental anatomy, see Chapter 15.

>> **Comparative anatomy:** The study of the similarities and differences among the anatomical structures of different species, including extinct species. Information from comparative anatomy can help scientists understand the human body's structures and processes. For example, comparing the anatomy of apes to that of humans shows us what particular structures allow for our ability to walk upright on two legs.

# A Little Chat about Jargon

Why does science have so many funny words? Why can't scientists just say what they mean, in plain English? Good question, with a two-part answer.

## Creating better communication

Scientists need to be able to communicate with others in their field. They say what they mean (most of them, most of the time, to the best of their ability), but what they mean can't be said in the English language that people use to talk about routine daily matters.

Like people working in every field, scientists develop vocabularies of technical terminology and other forms of jargon so they can better communicate with other scientists. It's important that the scientist sending the information and the scientist receiving it both use the same words to refer to the same phenomenon. To understand anatomy and physiology, you must know and use the same terminology, too. The jargon can be overwhelming at first, but understanding the reason for it and taking the time to learn it before diving into the complicated content will make your learning experience less painful.

## Establishing precise terminology

The second part of the answer starts with a little chat about jargon. Contrary to the belief of some, jargon is a good thing. *Jargon* is a set of words and phrases that people who know a lot about a particular subject use to talk together. There's jargon in every field (scientific or not), every workplace, every town, even every home. Families and close friends almost always use jargon in conversations with one another. Plumbers use jargon to communicate about plumbing. Anatomists and physiologists use jargon, much of which is shared with medicine and other fields of biology, especially human biology.

Scientists try to create terminology that's precise and easy to understand by developing it systematically. That is, they create new words by putting together existing and known elements. They use certain syllables or word fragments over and over to build new terms. With a little help from this book, you'll soon start to recognize some of these fragments. Then you can put the meanings of different fragments together and accurately guess the meaning of a term you've never seen before, just as you can understand a sentence you've never read before. Table 1-1 gets you started, listing some word fragments related to the organ systems we cover in this book.

**TABLE 1-1**     ## Technical Anatomical Word Fragments

| Body System | Root or Word Fragment | Meaning |
| --- | --- | --- |
| Skeletal system | os-, oste-; arth- | bone; joint |
| Muscular system | myo-, sarco- | muscle, striated muscle |
| Integument | derm- | skin |
| Nervous system | neur- | nerve |
| Endocrine system | aden-, estr- | gland, steroid |
| Cardiovascular system | card-, angi-, hema-, vaso- | heart (muscle), vessel, blood vessels |
| Respiratory system | pulmon-, bronch- | lung, windpipe |
| Digestive system | gastr-, enter-, dent-, hepat- | stomach, intestine, teeth, liver |
| Urinary system | ren-, neph-; ur- | kidney; urinary |
| Lymphatic system | lymph-, leuk-, -itis | lymph, white, inflammation |
| Reproductive system | andr-, uter- | male, uterine |

But why do these terms have to be Latin and Greek syllables and word fragments? Why should you have to dissect and put back together a term like *iliohypogastric?* Well, the terms that people use in common speech are understood slightly differently by different people, and the meanings are always undergoing change. Not so long ago, for example, no one speaking plain English used the term *laptop* to refer to a computer or *hybrid* to talk about a car. It's possible that, not many years from now, almost no one will understand what people mean by those words. Scientists, however, require consistency and preciseness to describe the things they talk about in a scientific context. The relative vagueness and changeability of terms in plain English makes this impossible. In contrast, Greek and Latin stopped changing centuries ago: *ilio, hypo,* and *gastro* have the same meaning now as they did 200 years ago.

**TIP**

Every time you come across an anatomical or physiological term that's new to you, see if you recognize any parts of it. Using this knowledge, go as far as you can in guessing the meaning of the whole term. After studying Table 1-1 and the other vocabulary lists in this chapter, you should be able to make some pretty good guesses.

# Looking at the Body from the Proper Perspective

Remember that story about a friend of a friend that went in to have a foot amputated only to awaken from surgery to find they removed the wrong one? This story highlights the need for a consistent perspective to go with the jargon. Terms that indicate direction make no sense if you're looking at the body the wrong way. You likely know your right from your left, but ignoring perspective can get you all mixed up. This section shows you the anatomical position, planes, regions, and cavities, as well as the main membranes that line the body and divide it into major sections.

## Getting in position

Stop reading for a minute and do the following: Stand up straight. Look forward. Let your arms hang down at your sides and turn your palms so they're facing forward. You are now in *anatomical position* (see Figure 1-1). Unless you are told otherwise, any reference to location (diagram or description) assumes this position. Using anatomical position as the standard removes confusion.

The following list of common anatomical descriptive terms (direction words) that appear throughout this and every other anatomy book may come in handy:

>> **Right:** Toward the patient's right

>> **Left:** Toward the patient's left

>> **Anterior/ventral:** Front, or toward the front of the body

>> **Posterior/dorsal:** Back, or toward the back of the body

>> **Medial:** Toward the middle of the body

>> **Lateral:** On the side or toward the side of the body

>> **Proximal:** Nearer to the point of attachment or the trunk of the body

» **Distal:** Farther from the point of attachment or the trunk of the body (think "distance")

» **Superficial:** Nearer to the surface of the body

» **Deep:** Farther from the surface of the body

» **Superior:** Above or higher than another part

» **Inferior:** Below or lower than another part

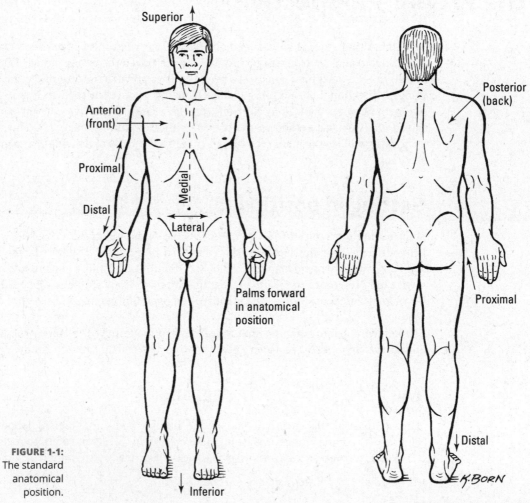

**FIGURE 1-1:**
The standard anatomical position.

Illustration by Kathryn Born, MA

Notice that this list of terms is actually a series of pairs. Learning them as pairs is more effective and useful.

## Dividing the anatomy

If you've taken geometry, you know that a *plane* is a flat surface and that a straight line can run between two points on that flat surface. Geometric planes can be positioned at any angle. In anatomy, generally three planes are used to separate the body into sections. Figure 1-2 shows you what each plane looks like. The reason for separating the body with imaginary lines — or making actual *cuts* referred to as *sections* — is so that you know which *half* or *portion* of the body or organ is being discussed. When identifying or comparing structures, you need to know your frame of reference. The anatomical planes are as follows:

>> **Frontal plane:** Divides the body or organ into anterior and posterior portions — think front and back.

>> **Sagittal plane:** Divides the body or organ lengthwise into right and left sections. If the vertical plane runs exactly down the middle of the body, it's referred to as the *midsagittal plane.*

>> **Transverse plane:** Divides the body or organ horizontally, into superior and inferior portions — think top and bottom. Diagrams from this perspective can be quite disorienting. You can think of the body like a music box that has a top that opens on a hinge. The transverse plane is where the music box top separates from the bottom of the box. Imagine that you open the box by lifting the lid and are looking down at the contents.

Anatomical planes do not always create two equal portions and can "pass through" the body at any angle. The three planes provide an important reference but don't expect the structures of the body, and especially the joints, to line up or move along the standard planes and axes.

## Mapping out your regions

The anatomical planes orient you to the human body, but *regions* (shown in Figure 1-3) compartmentalize it. Just like on a map, a region refers to a certain area. The body is divided into two major portions: axial and appendicular. The *axial body* runs right down the center (axis) and consists of everything except the limbs, meaning the head, neck, thorax (chest and back), abdomen, and pelvis. The *appendicular body* consists of appendages, otherwise known as *upper and lower extremities* (which you call arms and legs).

Frontal (coronal) plane

Midsagittal plane

Sagittal planes run parallel to the midsagittal plane.

Transverse (cross-sectional horizontal) plane

**FIGURE 1-2:** Planes of the body: frontal, sagittal, and transverse.

*Illustration by Kathryn Born, MA*

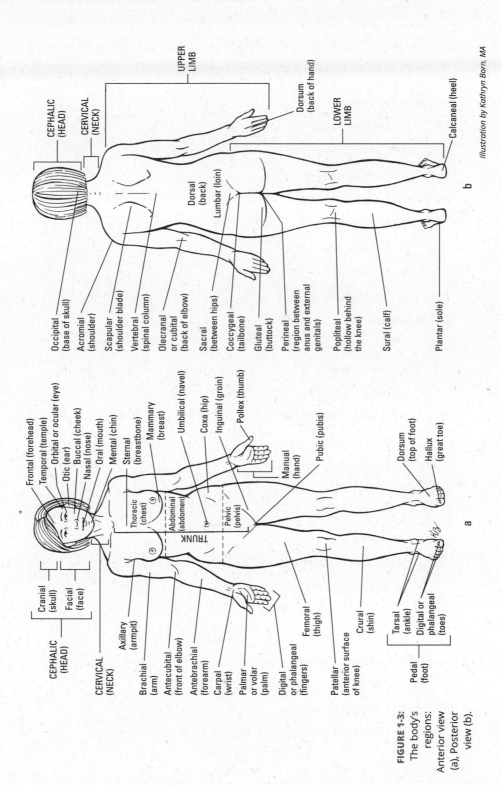

Illustration by Kathryn Born, MA

**FIGURE 1-3:**
The body's regions: Anterior view (a), Posterior view (b).

UPPER LIMB

CEPHALIC (HEAD)

CERVICAL (NECK)

Dorsum (back of hand)

LOWER LIMB

Calcaneal (heel)

Occipital (base of skull)
Acromial (shoulder)
Scapular (shoulder blade)
Vertebral (spinal column)
Olecranal or cubital (back of elbow)
Dorsal (back)
Lumbar (loin)
Sacral (between hips)
Coccygeal (tailbone)
Gluteal (buttock)
Perineal (region between anus and external genitals)
Popliteal (hollow behind the knee)
Sural (calf)
Plantar (sole)

b

Frontal (forehead)
Temporal (temple)
Orbital or ocular (eye)
Otic (ear)
Buccal (cheek)
Nasal (nose)
Oral (mouth)
Mental (chin)
Sternal (breastbone)
Mammary (breast)
Umbilical (navel)
Coxa (hip)
Inguinal (groin)
Pollex (thumb)
Manual (hand)
Pubic (pubis)
Dorsum (top of foot)
Hallux (great toe)

Cranial (skull)
Facial (face)
CEPHALIC (HEAD)
CERVICAL (NECK)
Axillary (armpit)
Brachial (arm)
Antecubital (front of elbow)
Antebrachial (forearm)
Carpal (wrist)
Palmar or volar (palm)
Digital or phalangeal (fingers)
Femoral (thigh)
Patellar (anterior surface of knee)
Crural (shin)
Tarsal (ankle)
Digital or phalangeal (toes)
Pedal (foot)

Thoracic (chest)
Abdominal (abdomen)
Pelvic (pelvis)
TRUNK

a

CHAPTER 1  Anatomy and Physiology: The Big Picture  17

Here's a list of the axial body's main regions:

» **Head and neck**
- Cephalic (head)
- Cervical (neck)
- Cranial (skull)
- Frontal (forehead)
- Nasal (nose)
- Occipital (base of skull)
- Oral (mouth)
- Orbital/ocular (eyes)

» **Thorax**
- Axillary (armpit)
- Costal (ribs)
- Deltoid (shoulder)
- Mammary (breast)
- Pectoral (chest)
- Scapular (shoulder blade)
- Sternal (breastbone)
- Vertebral (backbone)

» **Abdomen**
- Abdominal (abdomen)
- Gluteal (buttocks)
- Inguinal (bend of hip)
- Lumbar (lower back)
- Pelvic (area between hipbones)
- Perineal (area between anus and external genitalia)
- Pubic (genitals)
- Sacral (end of vertebral column)

Here's a list of the appendicular body's main regions:

» **Upper extremity**

- Antebrachial (forearm)
- Antecubital (inner elbow)
- Brachial (upper arm)
- Carpal (wrist)
- Cubital (elbow)
- Digital (fingers/toes)
- Manual (hand)
- Palmar (palm)

» **Lower extremity**

- Crural (shin, front of lower leg)
- Femoral (thigh)
- Patellar (front of knee)
- Pedal (foot)
- Plantar (arch of foot)
- Popliteal (back of knee)
- Sural (calf, back of lower leg)
- Tarsal (ankle)

# Casing your cavities

If you remove all the internal organs, the body is empty except for the bones and other tissues that form the space where the organs were. Just as a dental cavity is a hole in a tooth, the body's cavities are "holes" where organs are held (see Figure 1-4). The two main cavities are the *dorsal cavity* and the *ventral cavity.*

The dorsal cavity consists of two cavities that contain the central nervous system. The first is the *cranial cavity,* the space within the skull that holds your brain. The second is the *spinal cavity* (or *vertebral cavity*), the space within the vertebrae where the spinal cord runs through your body.

**Cranial cavity contains brain**

**Dorsal cavity**

**Pleural cavity**

**Thoracic cavity**

**Pericardial cavity**

**Diaphragm**

**Spinal cavity contains spinal cord**

**Ventral cavity**

**Abdominal cavity**

**Abdomino-pelvic cavity**

**Pelvic cavilty**

**FIGURE 1-4:**
The body's cavities.

*Illustration by Kathryn Born, MA*

The ventral cavity is much larger and contains all the organs not contained in the dorsal cavity. The ventral cavity is divided by the diaphragm into smaller cavities: the *thoracic cavity*, which contains the heart and lungs, and the *abdominopelvic cavity*, which contains the organs of the abdomen and the pelvis. The thoracic cavity is divided into the right and left *pleural cavities* (lungs) and the *pericardial cavity* (heart). The abdominopelvic cavity is also subdivided. The *abdominal cavity* contains organs such as the stomach, liver, spleen, and most of the intestines. The *pelvic cavity* contains the reproductive organs, the bladder, the rectum, and the lower portion of the intestines.

Additionally, the abdomen is divided into quadrants and regions. The mid-sagittal plane and a transverse plane intersect at an imaginary axis passing through the body at the *umbilicus* (navel or belly button). This axis divides the abdomen into *quadrants* (four sections). Putting an imaginary cross on the abdomen creates the right upper quadrant, left upper quadrant, right lower quadrant, and left lower quadrant. Physicians take note of these areas when a patient describes symptoms of abdominal pain.

The regions of the abdominopelvic cavity include the following:

>> **Epigastric:** The central part of the abdomen, just above the navel

>> **Hypochondriac:** Doesn't moan about every little ache and illness but lies to the right and left of the epigastric region and just below the cartilage of the rib cage (*chondral* means "cartilage," and *hypo-* means "below")

>> **Umbilical:** The area around the umbilicus

>> **Lumbar:** Forms the region of the lower back to the right and left of the umbilical region

>> **Hypogastric:** Below the stomach and in the central part of the abdomen, just below the navel

>> **Iliac:** Lies to the right and left of the hypogastric regions near the hipbones

# Organizing Yourself on Many Levels

Anatomy and physiology are concerned with the level of the individual body, what scientists call the *organism*. However, you can't merely focus on the whole and ignore the role of the parts. The life processes of the organism are built and maintained at several physical levels, which biologists call *levels of organization*: the cellular level, the tissue level, the organ level, the organ system level, and the organism level (see Figure 1-5). In this section, we review these levels, starting at the bottom.

Atoms

Molecules

Cells

Tissues

Organ

Organism

Organ System

Illustration by Kathryn Born, MA

**FIGURE 1-5:**
Levels of
organization in
the human body.

# Level I: The cellular level

If you examine a sample of any human tissue under a microscope, you see cells, possibly millions of them. All living things are made of cells. In fact, "having a cellular level of organization" is inherent in any definition of "organism." The work of the body actually occurs in the cells; for example, your whole heart beats to push blood around your body because of what happens inside the cells that create its walls.

# Level II: The tissue level

A *tissue* is a structure made of many cells — usually several different kinds of cells — that performs a specific function. Tissues are divided into four categories:

>> **Connective tissue** serves to support body parts and bind them together. Tissues as different as bone and blood are classified as connective tissue.

>> **Epithelial tissue (epithelium)** functions to line and cover organs as well as carry out absorption and secretion. The outer layer of the skin is made up of epithelial tissue.

>> **Muscle tissue** — surprise! — is found in the muscles, which allow your body parts to move; in the walls of hollow organs (such as intestines and blood vessels) to help move their contents along; and in the heart to move blood along via the acts of contraction and relaxation. (Find out more about muscles in Chapter 6.)

>> **Nervous tissue** transmits impulses and forms nerves. Brain tissue is nervous tissue. (We talk about the nervous system in Chapter 7.)

# Level III: The organ level

An *organ* is a group of tissues assembled to perform a specialized physiological function. For example, the stomach is an organ that has the specific physiological function of breaking down food. By definition, an organ is made up of at least two different tissue types; many organs contain tissues of all four types. Although we can name and describe all four tissue types that make up all organs, as we do in the preceding section, listing all the organs in the body wouldn't be so easy.

# Level IV: The organ system level

Human anatomists and physiologists have divided the human body into *organ systems*, groups of organs that work together to meet a major physiological need.

For example, the digestive system is one of the organ systems responsible for obtaining energy from the environment. Realize, though, that this is not a classification system for your organs. The organs that "belong" to one system can have functions integral to another system. The pancreas, for example, produces enzymes vital to the breakdown of our food (digestion), as well as hormones for the maintenance of our homeostasis (endocrine).

The chapter structure of this book is based on the definition of organ systems.

## Level V: The organism level

The whole enchilada. The real "you." As we study organ systems, organs, tissues, and cells, we're always looking at how they support you on the organism level.

## TAKING PICTURES OF YOUR INSIDES

For early anatomists like Hippocrates and da Vinci, the images they had were the sketches they made for themselves. The drawings made by Andreas Vesalius were compiled into the first anatomical atlas and the accuracy, considering it was the 16th century, is impressive. However, it is a German physicist named Wilhelm Conrad Roentgen who's remembered as "the father of medical imaging." In 1895, Roentgen changed the game by recording the first image of the internal parts of a living human: an X-ray image of his wife's hand. By 1900, X-rays were in widespread use for the early detection of tuberculosis, at that time a common cause of death. *X-rays* are beams of radiation emitted from a machine toward the patient's body, and X-ray images show details only of hard tissues, like bone, that reflect the radiation. In this way, they're similar to photographs. Refinements and enhancements of X-ray techniques were developed all through the 20th century, with extensive use and major advances during World War II. The X-ray is still a widely used method for medical diagnosis, not just for bone breaks but for screening for signs of disease, especially tumors.

In the 1970s, computer technology took off, taking medical imaging technology with it. Digital imaging techniques began to be applied to convert multiple flat-slice images into one three-dimensional image. The first technology of this sort was called *computed axial tomography* (commonly called a CAT or CT scan). The technique combines multiple X-ray images of varying depths into images of whole structures inside the body. Contrast dye can be used to highlight particular areas, which is especially useful for a quick assessment (for example, after a trauma).

Another class of imaging technology utilizing radiation is positron emission tomography (PET). A radioactive isotope can be attached to a specific molecule — a drug, for example. After administering the drug to the patient, the isotope emits radiation, which can be traced and followed with radiation detectors. This is especially useful for testing the efficacy of drugs in a clinical research setting. It's unique in that the scan provides information of organ function on a cellular level.

*Ultrasound* imaging technology uses the echoes of sound waves sent into the body to generate a signal that a computer turns into a real-time image of anatomy and physiology. Ultrasound can also produce audible sounds, so the anatomist or physiologist can, for example, watch the pulsations of an artery while hearing the sound of the blood flowing through it. Although all these technologies are considered noninvasive, ultrasound is the least invasive of all (no radiation) so it's used more freely, especially in sensitive situations like pregnancy.

*Magnetic resonance imaging* (MRI) utilizes magnetic fields and radio pulses to create an image of the interior. Soft tissue structures are more difficult to scan using other methods, especially those found underneath bone. The resulting 3D image can pinpoint anomalies within an organ, often in great detail. Since the early 1990s, neuroscientists have been using a type of specialized MRI scan, called *functional MRI* (fMRI), to acquire images of the brain. Images can be recorded over time, and the active areas of the brain "light up" on the scan, showing which parts are active during specific tasks. Basically, fMRI enables scientists to watch a patient's or research subject's thoughts as he or she is thinking them!

Digital imaging technologies produce images that are extremely clear and detailed. The images can be produced much more quickly and cheaply than older technologies allowed for, and the images can be easily duplicated, transmitted, and stored. The amount of anatomical and physiological knowledge that digital imaging technologies have helped generate over the past 30 years has transformed biological and medical science. As the techniques are continually researched and developed, our understanding of physiology and accuracy in diagnostics will continue to improve.

Chapter **2**

# What Your Body Does All Day

This chapter is about your life as an organism. As Chapter 1 explains, *organism* is the fifth of five levels of organization in living things. Although the word *organism* has many possible definitions, for the purposes of this chapter, an organism is a living unit that metabolizes and maintains its own existence.

In this chapter, you see why your to-do list, crowded as it is, doesn't include items such as *Take ten breaths every minute* or *At 11:30 a.m., open sweat glands.* The processes that your body must carry out minute by minute to sustain life, not to mention the biochemical reactions that happen millions of times a second, can't be left to the distractible frontal lobes (the conscious, planning part of your brain). Instead, your organs and organ systems function together smoothly to carry out these processes and reactions automatically, without the activity ever coming to your conscious attention. All day and all night, year in and year out, your body builds, maintains, and sustains every part of you; keeps your temperature and your fluid content within some fairly precisely defined ranges; and transfers substances from outside itself to inside, and then back out again. These are the processes of *metabolism* and *homeostasis*.

# Transferring Energy: A Body's Place in the World

The laws of thermodynamics are the foundation of how the physics and chemistry of the universe are understood. They're at the "we hold these truths to be self-evident" level for chemists and physicists of all specialties, including all biologists. The first law of thermodynamics states that energy can be neither created nor destroyed — it can only change form. (Turn to Chapter 16 for a brief look at the first law and other basic laws of chemistry and physics.) Energy changes form continuously — within stars, within engines of all kinds, and, in some very special ways, within organisms.

The most basic function of the organism that is you on this planet is to take part in this continuous flow of energy. As a *heterotroph* (an organism that doesn't photosynthesize), you ingest (take in) energy in the form of matter — that is, you eat the bodies of other organisms. You use the energy stored in the chemical bonds of that matter to fuel the processes of your *metabolism* and *homeostasis.* That energy is thereby transformed into matter called "you" (the material in your cells), matter that's "not you" (the material in your exhaled breath and in your urine), and some heat radiated from your body to the environment.

**TECHNICAL STUFF**

*Hetero* means "other," and *tropho* means "nourishment." A *heterotroph* gets its nourishment from others, as opposed to an *autotroph*, which makes its own nourishment, as a plant does.

Plants convert light energy from the sun into the chemical energy in carbohydrates, which comprise most of the matter of the plant bodies, recycling the waste matter (carbon dioxide) of your metabolic processes. Energy goes around and around, and some of it is always flowing through your body, being transformed constantly as it does so. You, my friend, are part of a cycle of cosmic dimensions!

# Building Up and Breaking Down: Metabolism

The word *metabolism* describes all the chemical reactions that happen in the body. These reactions are of two kinds — *anabolic reactions* make things (molecules), and *catabolic reactions* break things down.

**TIP**

To keep the meanings of anabolic and catabolic clear in your mind, associate the word *catabolic* with the word *catastrophic* to remember that catabolic reactions break down products. Then you'll know that anabolic reactions create products.

Your body performs both anabolic and catabolic reactions at the same time, around the clock, to keep you alive and functioning. Even when you're sleeping, your cells are busy. You just never get to rest (until you're dead).

Chapter 11 gives you the details on how the digestive system breaks down food into nutrients and gets them into your bloodstream. Chapter 9 explains how the bloodstream carries nutrients around the body to every cell and carries waste products to the urinary system. Chapter 12 shows you how the urinary system filters the blood and removes waste from the body. This chapter describes the reactions that your cells undergo to convert fuel to usable energy. Ready?

## Why your cells metabolize

Even when your outside is staying still, your insides are moving. Day and night, your muscles twitch and contract and maintain "tone." Your heart beats. Your blood circulates. Your diaphragm moves up and down with every breath. Nervous impulses travel. Your brain keeps tabs on everything. You think. Even when you're asleep, you dream (a form of thinking). Your intestines push the food you ate hours ago along your alimentary canal. Your kidneys filter your blood and make urine. Your sweat glands open and close. Your eyes blink, and even during sleep, they move. Men produce sperm. Women move through the menstrual cycle. The processes that keep you alive are always active.

Every cell in your body is like a tiny factory, converting raw materials to useful molecules such as proteins and thousands of other products, many of which we discuss throughout this book. The raw materials (nutrients) come from the food you eat, and the cells use the nutrients in metabolic reactions. During these reactions, some of the energy from catabolized nutrients is used to generate a compound called *adenosine triphosphate* (ATP). This molecule is the one your cells can actually use to power all those chemical reactions.

**REMEMBER**

So, nutrients are catabolized (broken down), ATP is formed (anabolized), and when needed, ATP is catabolized (for energy). This principle of linked anabolic and catabolic reactions is one of the cornerstones of human physiology and is required to maintain life. Cellular metabolism also makes waste products that must be removed (exported) from the cell and ultimately from the body.

ATP works like a rechargeable battery. It contains three phosphates aligned in a row (see Figure 2-1). Breaking one of them off accesses the energy, leaving behind *adenosine diphosphate* (ADP) and a phosphate (P) by itself. However, just like you can plug in your phone to recharge its battery, the energy in the bonds of glucose is used to reattach the P — re-creating ATP (albeit in an incredibly complicated way).

**FIGURE 2-1:** The chemical structure of ADP and ATP.

## How your cells metabolize

The reactions that convert fuel (specifically glucose) to usable energy (ATP molecules) include glycolysis, the Krebs cycle (aerobic respiration) and anaerobic respiration, and oxidative phosphorylation. Together, these reactions are referred to as *cellular respiration*. These are complex pathways, so expect to take some time to understand them. See Figure 2-2 and refer to it as many times as necessary to understand what happens in cellular respiration. (*Note:* Alcohol fermentation is included for reference but does not occur in the human body.)

Here, we focus on the ins and outs of the three main components of cellular respiration.

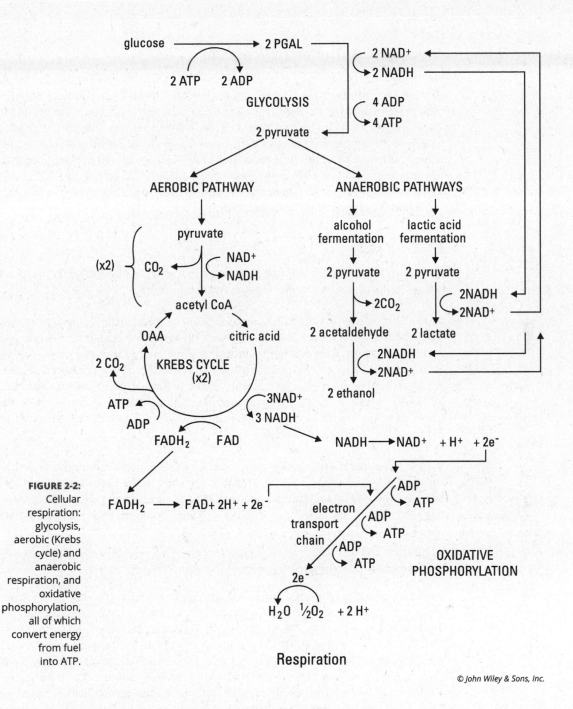

**FIGURE 2-2:** Cellular respiration: glycolysis, aerobic (Krebs cycle) and anaerobic respiration, and oxidative phosphorylation, all of which convert energy from fuel into ATP.

Respiration

© John Wiley & Sons, Inc.

## Glycolytic pathway (glycolysis)

Starting at the top of Figure 2-2, you can see that glucose — the smallest molecule that a carbohydrate can be broken into during digestion — goes through the process of *glycolysis*, which starts cellular respiration and uses some energy (ATP) itself. Glycolysis occurs in the cytoplasm and doesn't require oxygen. Two molecules of ATP are required to start each molecule of glucose rolling down the glycolytic pathway; although four molecules of ATP are generated during glycolysis, the net production of ATP is two molecules. In addition to the two ATPs, two molecules of *pyruvic acid* (also called *pyruvate*) are generated. They move into a mitochondrion and enter the Krebs cycle.

## Krebs cycle

The *Krebs cycle* is a major biological pathway in the metabolism of every multicellular organism. It's an *aerobic pathway*, requiring oxygen.

As the pyruvate enters the mitochondrion, a molecule called *nicotinamide adenine dinucleotide* (NAD+) joins it. NAD+ is an electron carrier (that is, it carries energy), and it gets the process moving by bringing some energy into the pathway. The NAD+ provides enough energy that when it joins with pyruvate, carbon dioxide is released, and the high-energy molecule NADH is formed. *Flavin adenine dinucleotide* (FAD) works in much the same way, becoming $FADH_2$. The product of the overall reaction is *acetyl coenzyme A* (acetyl CoA), which is a carbohydrate molecule that puts the Krebs cycle in motion.

**REMEMBER**

Cycles are endless. Products of some reactions in the cycle are used to keep the cycle going. An example is acetyl CoA: It's a product of the Krebs cycle, yet it also helps initiate the cycle. With the addition of water and acetyl CoA, *oxaloacetic acid* (OAA) is converted to *citric acid.* Then, a series of reactions proceeds throughout the cycle.

## Oxidative phosphorylation

Oxidative phosphorylation, also called the *electron transport chain* (ETC), takes place in the inner membrane of the mitochondria. The electron carriers produced during the Krebs cycle — NADH and $FADH_2$ — are created when NAD+ and FAD, respectively, are "reduced." When a substance is *reduced,* it gains electrons; when it's *oxidized,* it loses electrons. (Turn to Chapter 16 for more information about such "redox reactions.") So NADH and $FADH_2$ are compounds that have gained electrons, and therefore, energy. In the ETC, oxidation and reduction reactions occur repeatedly as a way of transporting energy. At the end of the chain, oxygen atoms accept the electrons, producing water. (Water from metabolic reactions isn't a significant contributor to the water needs of the body.)

As NADH and FADH$_2$ pass down the respiratory (or electron transport) chain, they lose energy as they become oxidized and reduced, oxidized and reduced, oxidized and . . . . It sounds exhausting, doesn't it? Well, their energy supplies become exhausted for a good cause. The energy that these electron carriers lose is used to add a molecule of phosphate to adenosine *di*phosphate (ADP) to make it adenosine *tri*phosphate — the coveted ATP. For each NADH molecule that's produced in the Krebs cycle, three molecules of ATP can be generated. For each molecule of FADH$_2$ that's produced in the Krebs cycle, two molecules of ATP are made.

**REMEMBER**

Theoretically, the entire process of aerobic *cellular respiration* — glycolysis, Krebs cycle, and oxidative phosphorylation — generates a total of 38 ATP molecules from the energy in one molecule of glucose: 2 from glycolysis, 2 from the Krebs cycle, and 34 from oxidative phosphorylation. However, this theoretical yield is never quite reached because processes, especially biological processes, are never 100 percent efficient. In the real world, usually around 29 to 30 ATP molecules per glucose molecule are expected.

### Anaerobic respiration

Sometimes oxygen isn't present, but your body still needs energy. During these times, a backup system, an *anaerobic pathway* (called anaerobic because it proceeds in the absence of oxygen) exists. Lactic acid fermentation generates NAD$^+$ so that glycolysis, which results in the net production of two molecules of ATP, can continue. However, if the supply of NAD$^+$ runs out, glycolysis can't occur, and ATP can't be generated.

This occurs most often in muscle cells during periods of intense exercise. The byproduct of this reaction, *lactic acid,* builds up in the muscle, contributing to *muscle fatigue* (the inability of a muscle cell to contract). Thus, this process cannot be maintained for extended periods of time.

# Staying in Range: Homeostasis

Chemical reactions are not random events. Any reaction takes place only when all the conditions are right for it: All the required reagents and catalysts are close together in the right quantities; the fuel for the reaction is present, in sufficient amount and in the right form; and the environmental variables are all within the right range, including the temperature, salinity, and pH. The complicated chemistry of life is extremely sensitive to the environmental conditions; the environment being the body itself. *Homeostasis* is the term physiologists use to reference the balance of all the variables. Numerous homeostatic mechanisms are employed by our bodies to keep everything in check; otherwise, all the reactions that comprise our metabolism cannot occur.

The following sections look at a few important physiological variables and how the mechanisms of homeostasis keep them in the optimum range in common, everyday situations.

As metabolic reactions, homeostatic reactions require energy.

# Maintaining a constant temperature: Thermoregulation

All metabolic reactions in all organisms require that the temperature of the body be within a certain range. Because we humans are *homeotherms* or "warm-blooded," we maintain a relatively constant body temperature regardless of the ambient temperature. We do this by regulating our metabolic rate. The large number of mitochondria per cell enables a high rate of metabolism, which generates a lot of heat.

Regulating body temperature requires a steady supply of fuel (glucose) to the mitochondrial furnaces.

Another way we control our body temperature is by employing adaptations that conserve the heat generated by metabolism within the body in cold conditions or dissipate that heat out of the body in overly warm conditions. A few of the specific adaptations include the following:

>> **Sweating:** Sweat glands in the skin open their pores to dissipate heat by evaporative cooling of water from the skin. They close to conserve heat. Sweat glands are opened and closed by the action of muscles at the base of the gland, deep under the skin. Refer to Chapter 4.

>> **Blood circulation:** Blood vessels close to the skin dilate (enlarge) to dissipate heat in the blood through the skin. They constrict (narrow) to conserve heat. That's why your skin flushes (reddens) when you're hot: That's the color of your blood visible at the surface of your skin. Refer to Chapter 9.

>> **Muscle contraction:** When closing sweat pores and blood vessel constriction are not enough to conserve heat in cold conditions, your muscles will begin to contract automatically to generate more heat. This reaction is familiar as "shivering."

>> **Insulation:** Regions of fatty tissue under the skin provide insulation, holding the warmth of the body in. Body hair aids in this, too (though not enough to keep us from needing a nice, warm winter coat).

# Swimming in H₂O: Fluid balance

A watery environment is a requirement for a great proportion of metabolic reactions (the rest need a lipid, or fatty, environment). The body contains a lot of water: in your blood, in your cells, in the spaces between your cells, in your digestive organs, here, there, and everywhere. Not pure water, though. The water in your body is a solvent for thousands of different ions and molecules (solutes). The quantity and quality of the solutes change the character of the solution. Because solutes are constantly entering and leaving the solution as they participate in or are generated by metabolic reactions, the characteristics of the watery solution must remain within certain bounds for the reactions to continue happening.

>> **Changes in the composition of urine:** The kidney is a complex organ that has the ability to measure the concentration of many solutes in the blood, including sodium, potassium, and calcium. Very importantly, the kidney can measure the volume of water in the body by sensing the pressure of the blood as it flows through (the greater the volume of water, the higher the blood pressure). If changes must be made to bring the volume and composition of the blood back into the ideal range, the various structures of the kidney incorporate more or less water, sodium, potassium, and so on into the urine. That's why your urine is paler or darker at different times. This and other functions of the urinary system are discussed in Chapter 12.

>> **The thirst reflex:** Water passes through your body constantly: mainly, in through your mouth and out through various organ systems, including the skin, the digestive system, and the urinary system. If the volume of water falls below the optimum level (dehydration), and the kidneys alone can't regain the balance, the mechanisms of homeostasis intrude on your conscious brain to make you uncomfortable. You feel thirsty. You ingest something watery. Your fluid balance is restored and your thirst reflex leaves you alone.

# Adjusting the fuel supply: Blood glucose concentration

Glucose, the fuel of all cellular processes, is distributed to all cells dissolved in the blood. The concentration of glucose in the blood must be high enough to ensure that the cells have enough fuel. However, extra glucose beyond the immediate needs of the cells can harm many important organs and tissues, especially where the vessels are tiny, as in the retina of the eye, the extremities (hands and, especially, feet), and the kidneys. Diabetes is a disease in which there is a chronic overconcentration of glucose in the blood.

The amount of glucose in the blood is controlled mainly by the pancreas. Absorption by the small intestine puts the glucose from ingested food into the blood.

Insulin is a hormone released into the blood from the pancreas in response to increased blood glucose levels. Most cells have receptors that bind the insulin, which allows glucose into the cells for cellular respiration. The cells of the liver, muscles, and adipose tissue (fat) take up the glucose and store it as glycogen (see Chapter 3). At times when your intestines aren't absorbing much glucose, like hours after a meal, the production of insulin is suppressed and the stored glucose is released into the blood again. Refer to Chapter 8 for more information about the pancreatic control of blood glucose levels.

## Measuring important variables

How does the pancreas know when to release insulin and how much is enough? How does the kidney know when the salt content of the blood is too high or the volume of the blood is too low? What tells the sweat glands to open and close to cool the body or retain heat? Well, read on!

## FEEDBACK IN PHYSIOLOGY

In biology and other sciences, *feedback* is information that a system generates about itself or its effects that influences how its processes continue. Feedback mechanisms can be *negative* or *positive*. These terms do not mean that one is harmful and the other beneficial, and they are not opposites — that is, they do not counteract one another in the same system or process. Organisms use both types of feedback mechanisms to control different aspects of their physiology.

A *negative feedback mechanism* functions to keep things within a range. It tells the system to stop, slow down, decrease its output when the optimum quantity or range has been achieved or to speed up or increase its output when the quantity is below the optimum range. In other words, it tells a process to start doing the opposite of what it's doing now. Negative feedback mechanisms maintain or regulate physiological conditions within a set and narrow range. Homeostasis depends on a vast array of negative feedback mechanisms.

A *positive feedback mechanism* tells a process to continue or increase its output. Positive feedback says, "Some is good; more would be better." It accelerates or enhances the output created by a stimulus that has already been activated. A positive feedback mechanism is usually a cascading process that increases the effect of the stimulus and pushes levels out of normal ranges, usually for a specific and temporary purpose. Because positive feedback can get out of control (think fire), there are relatively few positive feedback mechanisms. One example is the "clotting cascade" that occurs in response to a cut in a blood vessel, described in this chapter. Another is the release of oxytocin to intensify uterine contractions during childbirth (described in Chapter 14).

Every homeostatic mechanism employs three parts: a receptor, an integrator, and an effector. Numerous *receptors*, or sensors, are strategically placed throughout your body. Some respond to chemical changes (like pH), others to mechanical ones (like blood pressure), and there are many more. These receptors are specialized nervous cells and communicate to the brain — the *integrator* — any changes from our balance. The brain processes all the incoming information and "decides" if a response is warranted. If it is, a message will be sent out via neurons or hormones to the *effectors*, which carry out the body's response (the effect).

# Growing, Replacing, and Renewing

My, how you've changed, and are still changing! Growing up, growing old, and just living every day, you're building new parts and replacing old ones. From conception to early adulthood, your body was busy making itself: everything from scratch.

But the job wasn't finished when you were fully grown. Complex living tissues and organs almost all require replacement parts at some time, and many require them all the time. This necessity is one of the defining characteristics of organisms — the ability to organize matter into the structures that compose themselves and to replace and renew those structures as required, as we describe in the following sections.

As we discuss in the "Building Up and Breaking Down: Metabolism" section earlier in the chapter, making new cells and tissues is *anabolic metabolism*, and breaking up and eliminating old cells and tissues is *catabolic metabolism*.

## Growing

You began life as a single cell and built yourself from there, with some help from your mom to get started. Your body developed along a plan, building a backbone with a head at the top and a tail at the bottom (somehow, you lost the tail). Now look at you: 100 trillion cells, almost every one with its own special structure and job to do. Good work! Find more about the processes of development in Chapter 15.

## Replacing

Just like the organism they're a part of, many kinds of cells have a life cycle: They're born, they develop, they work, they get worn out, and they die. For an organism to continue its life cycle, these cells must be replaced continuously, either by the division of the same cell type or by the differentiation of *stem cells*. These relatively undifferentiated cells wait patiently until they're called upon to divide. Some of the daughter cells differentiate into their specific programmed type while others remain stem cells and wait to be called upon the next time. Stem cells are an active area of research in physiology and in the field of regenerative medicine.

Some of the cell and tissue types that must be continuously replaced are

>> **Red blood cells:** The life cycle of a red blood cell is about 120 days. That means you replace all your red blood cells three times a year. New ones come from the red marrow of the bones, and old ones are scavenged for iron in the spleen and then broken down in the liver.

>> **Epidermis:** The cells of the epidermis, the outer layer of the skin, are constantly shed from the surface and replaced from below. Your body replaces the entire epidermis about every six weeks. This process is discussed in Chapter 4.

>> **Intestinal lining:** The epithelial cells of the intestinal lining are replaced about every week. You realize what a feat this is when you read about the intestine in Chapter 11.

>> **Respiratory membrane:** Your body replaces the epithelial cells that line the alveolar wall and the pulmonary capillary vessels about every week. Refer to Chapter 10 for a description of the respiratory membrane.

>> **Sperm:** The process of *spermatogenesis* (making sperm) is continuous, beginning in a male's puberty and ending with his death. The quantity and quality varies with his age and health. Turn to Chapter 14 for more details.

>> **Bone:** As you can read about in Chapter 5, bone is living tissue and very active in a number of ways. The bones bear the body's weight and the stress of impact. Tiny cracks develop in bone all the time and are repaired quickly and constantly, a process known as *remodeling*. Bones serve as a storage depot for metal ions, especially calcium, which flows in and out of the bone constantly.

Some other types of tissue replace their cells at a very slow rate, such as the following:

>> **Brain cells:** Scientists thought for many decades that brain cells that died weren't replaced, and, in general, that no new cells were developed in the brain during adulthood. Brain researchers have now shown that this isn't true.

Still, neurons are designed to not need replacement, they're meant to last you a lifetime. The processes whereby new cells are born in the adult brain have attracted much research interest. See Chapter 7.

>> **Cardiac muscle:** Until recently, physiologists believed that cardiac muscle cells couldn't regenerate, but that belief has recently been called into question. In 2009, researchers in Sweden reported evidence that, in healthy hearts, cardiac muscle cells do indeed divide, but slowly. The researchers estimated that a 20-year-old renews about 1 percent of heart muscle cells per year and that about 45 percent of the cardiac muscle cells of a 50-year-old are generated after birth. Research published in the early 2000s showed evidence that cardiac muscle cells regenerate to some extent after a heart attack.

## Repairing parts

Your body repairs some tissues as necessary, such as after an injury:

>> **Skeletal muscle:** Mature skeletal muscle cells, called *fibers,* don't divide and aren't replaced unless they're damaged. After they're formed, skeletal muscle fibers generally survive for your entire lifetime. But wait, you say that you've been working out and your biceps are twice as big as they were last year? Congratulations, but you didn't add cells. The cells you had just got bigger.

>> **Smooth muscle:** Like skeletal muscle fibers, smooth muscle fibers are replaced when they're injured.

>> **Skin fibroblasts:** These are different from epidermal cells. These cells proliferate rapidly to repair damage from a cut or wound and are responsible for generating scar tissue, as discussed in the next section.

>> **Liver cells:** Normally, these cells divide only rarely. However, if large numbers of liver cells are removed — by surgical removal of part of the liver, for example — the remaining cells proliferate rapidly to replace the missing tissue. This makes it possible to transplant part of the liver of a living donor to a recipient, or to split a single liver from a nonliving donor to two recipients. In these cases, when all goes well, both parts regenerate into a complete and functioning liver.

## Healing wounds

When you have a tiny, superficial surface wound (a little scratch), the epidermis simply replaces the damaged cells. In a few days, the scratch is gone. But when the wound is deep enough that blood vessels are damaged, the healing process is a little bit more involved. Turn to Chapter 9 for information on blood and blood vessels.

The immediate rush of blood washes debris and microbes out of the wound. Then, the vessels around the wound constrict to slow down the blood flow. A type of "formed element" in the blood called *platelets* sticks to the collagen fibers that make up the vessel wall, forming a natural band–aid called a *platelet plug.*

After the platelet plug forms, a complex chain of events results in the formation of a *clot* that stops blood loss altogether. This chain of events is called the *clotting cascade* or *coagulation cascade.* Enzymes called *clotting factors* initiate the cascade. Here's a rundown of what happens, focusing on the most important steps:

>> **Prothrombin:** This clotting factor converts to thrombin. Calcium is required for this reaction.

>> **Thrombin:** This factor acts as an enzyme and causes the plasma protein *fibrinogen* to form long threads called *fibrin.*

>> **Fibrin threads:** Wrapping around the platelet plug, these threads form a meshlike template for a clot.

>> **Clot:** The meshlike structure traps the red blood cells and forms a clot. As the red blood cells that are trapped on the outside of the clot dry out (or the air oxidizes the iron in them, like rust), they turn a brownish-red color, and a scab forms.

Underneath the scab, the blood vessels regenerate and repair themselves, and in the dermis, cells called *fibroblasts* spur the creation of proteins to fill the space in the damaged layers. Scars are created to provide extra strength to skin areas that are deeply wounded. Scar tissue has many interwoven collagen fibers, but no hair follicles, nails, or glands. Feeling maybe lost in the area covered with scar tissue if the nerves are damaged.

## Lasting parts

As we mention earlier in the chapter, *almost* all tissues and organs require replacement parts at some time. However, here are some exceptions:

>> **Central nervous system:** For the most part, the cells and tissues of the central nervous system are incapable of self-repair and regeneration. Thus the poor prognosis in cases of spinal cord injury.

» **Peripheral nerves:** These are the nerve cells that transmit sensation or motor messages between the central nervous system and the skin and skeletal muscles (see Chapter 7). Many types of peripheral neurons don't undergo regular replacement in normal functioning. They are therefore some of the oldest cells in your body. Unfortunately, they're not regenerated when they die from injury, so some kinds of nerve damage are permanent. Because they're not replaced when they die, the number of such nerve cells declines throughout life.

» **Ova:** A woman has all the eggs she's ever going to have in her ovaries at birth. In most women, that's about half a million more than they'll ever need. Most eggs die before puberty. Only a few mature and participate in the monthly events of the ovarian (menstrual) cycle. And only a very, very few go on to participate in the events of reproduction described in Chapter 14.

# Chapter **3**

# A Bit about Cell Biology

B iologists see life as existing at five "levels of organization," of which the cellular level is the first (see Chapter 1 for more on the levels of organization). A basic principle of biology says that all organisms are made up of cells and that anything that has even one cell is an organism. Understanding the basics of cell biology is necessary for understanding any aspect of biology, including human anatomy and physiology.

The first thing to know is that cell biology is astoundingly complex. You can gain a detailed understanding of this complexity only from years of hard study. This chapter aims only to give you some idea of how complex cell biology is, so as to provide context for the various physiological miracles we describe in later chapters.

## The Functions of Cells

Almost all the structures of anatomy are built of cells, and almost all the functions of physiology are carried out within cells. A comprehensive list of cell functions would be impossible, but we can group cell functions into a few main categories, which we do in the following sections.

# Building themselves

Cells arise from other cells and nowhere else. Once in an organism's lifetime, at the beginning, two cells fuse to form a new cell. Ever after in that organism's lifetime, two cells arise from the division of one cell, and all the cells ultimately derive from the first one. This process is how an organism builds itself from one single generic cell, called the *zygote,* to a complex organism comprising trillions of highly differentiated, highly specialized, and highly efficient cells all working together in a coordinated way. Here's a look at how a cell goes from one to many.

## Fusing: The zygote

The organism's first cell is the *zygote,* made by the fusion of sex cells: an *ovum* (egg cell) from the female parent and a *sperm* cell from the male parent. (See Chapter 14 for more information about sex cells, and see Chapter 15 for more about the zygote.) The zygote has two complete copies of DNA: one from its male parent and one from its female parent. The two copies combine in the zygote's nucleus. The zygote is said to be *diploid* — having a complete double-set of DNA.

## Dividing: Mitosis

In the form of cell division called *mitosis,* one cell divides into two *daughter cells,* each of them complete but smaller than the original cell. Details of this process are discussed later in this chapter.

**TECHNICAL STUFF**

The process of mitosis occurs only in diploid cells — cells that have two copies of the DNA. That is, all cells except the mature sex cells, which are *haploid* — they have only one copy of the organism's DNA.

## Differentiating

After mitosis is complete, each daughter cell goes on to its own separate life. One or both may start or continue down a path of *differentiation,* the name for processes that give cells their particular structures and functions. A cell destined to become a nerve cell starts down one path of differentiation; a cell destined to become a muscle cell starts down another path.

A variation on this mechanism involves a special kind of cell called a *stem cell.* A stem cell divides by mitosis, and one daughter cell remains a stem cell and goes on dividing again and again, while the other daughter cell goes on to differentiate into a specific type of cell in a particular tissue. Only some tissues have their own special stem cells, such as the skin and the blood.

The details of cellular differentiation are beyond the scope of this book. Its complexity is beyond imagining. This is the only explanation you really need: It's under genetic control.

# Building tissues

All tissues are made of cells, and those cells build and maintain it. Cells in a tissue are to one degree or another *differentiated* or *specialized* for their anatomical or physiological function in the tissue.

In addition to the cells, many tissues also contain structural proteins (which are made by the cells). Differentiated cells produce different proteins: Some produce only a few different proteins, and some produce many different proteins in response to signals they receive from other cells. The process of protein construction is basically the same in every cell and for every protein.

REMEMBER

For the purposes of anatomy and physiology, remember that all cells have certain very important features in common, but they differentiate into a vast array of shapes and sizes, containing vastly diverse structures, and having different functions and life cycles.

# Transforming energy

Most cells make ATP to fuel their own metabolism. They use glucose in the process of cellular respiration to do so. Flip to Chapter 2 for a description of how cells make ATP.

Some cells, like those lining the small intestine, absorb the glucose only to send it out the other side — allowing other cells access to this valuable resource. Sometimes, the glucose made available by the digestive system is more than the body can use at that moment. Thus, some cells, like those in the liver, function to corral the extra glucose molecules and store them. Later, when glucose levels are low, these cells release some of their stores, making it available to other cells.

# Making and transporting products

A great many types of cells make special chemicals that are incorporated into tissues and participate in metabolic reactions. Cellular products include thousands of specific proteins and polypeptides, signaling chemicals like neurotransmitters and hormones, small molecules and ions, lipids of many kinds, and structural molecules of many kinds.

Some specialized cells do essentially nothing else but make and export one product for use by other cells; others make products and perform other functions.

Some cells specialize in transporting the products of other cells around the body, or in transporting metabolic waste products out of the body. Some of these transporting cells have other functions as well. Others do nothing but that one job through their entire life cycle. Red blood cells are an extreme example of the one-job model. They lose their nuclei during differentiation and thereafter do nothing but transport gas molecules from one place to another. They don't divide, don't produce ATP, and don't maintain themselves. When their gas-transporting structures wear out, RBCs have nothing to do. They're removed from circulation and broken down in the liver.

## Communicating

Some cells transmit signals of various kinds while remaining in one place in the body. Some nervous cells' sole purpose is to generate and conduct electrical signals. They typically live for years, often until the death of the organism itself. Other cells produce various kinds of signaling molecules, like hormones and neurotransmitters, or receive and react to those signaling molecules.

# Seeing the Inside of Eukaryotic Cells

Although they're astoundingly varied, cells are also remarkably alike. (This is a theme of cell biology.) It's not just that all the cells of one organism or even one species are a lot alike. All cells, at least all *eukaryotic* cells, are alike. Plants, animals, and fungi are *eukaryotes* (organisms made up of eukaryotic cells), and all their cells, in all their enormous complexity and variation, are fundamentally alike. Yes, your skin cells, your kidney cells, and your bone cells are fundamentally similar to the leaf cells and root cells of a carrot; the cells of a mold, mushroom, or yeast; and the single cell of microorganisms called *protists* that live in water and soil.

Here's a simplistic description of a eukaryotic cell: It's a membrane-bound sac containing smaller but distinctive structures, called *organelles* ("little organs"), suspended in a gel-like matrix called the *cytoplasm.* As their name suggests, organelles are functional subunits of a cell, as organs are functional subunits of an organism. One of the largest and most prominent organelles, the *nucleus,* controls a cell's functioning, similar to the way the nervous system controls an organism's functioning. The term *eukaryote* is derived from the Greek term *karyos,* meaning "nut" or "kernel," which early biologists used to refer to the nucleus. Figure 3-1 shows the general structure of a eukaryotic cell. Refer to this figure as you read about the various cellular structures in the following sections. Table 3-1 gives you an overview of the structures found within a eukaryotic cell.

Cytoplasm

Cilia

Vesicle formation

Nucleolus

Nucleus

Mitochondrion

Vacuole

Ribosomes

Rough endoplasmic reticulum

Smooth endoplasmic reticulum

Golgi apparatus

Lysosome

Centriole

**FIGURE 3-1:** A cutaway view of a basic animal cell and its organelles.

Plasma membrane

*Illustration by Kathryn Born, MA*

## TABLE 3-1     Organelles of Animal Cells (Including Humans)

| Organelle | Function |
| --- | --- |
| Nucleus | Controls the cell; houses the genetic material |
| Mitochondrion | Cell "powerhouse"; site of cellular respiration |
| Endoplasmic reticulum | Plays an important role in protein synthesis; participates in transporting cell products; involved in metabolizing (breaking down) fats as well as drugs |
| Ribosome | Binds amino acids together under the direction of mRNA to make protein |
| Golgi apparatus | Modifies proteins into functional form; "packages" cellular products in sacs called *vesicles* in which products can cross the cell membrane to exit the cell |
| Vacuoles | Membrane-bound spaces in the cytoplasm that aid in endo- and exocytosis |
| Lysosomes | Contain enzymes that break down harmful cell products and waste materials and actively transport them out of the cell |

**TECHNICAL STUFF**

The organisms called *bacteria* (singular, *bacterium*) are made up of *prokaryotic* cells. Prokaryotic cells are very different from and much simpler than eukaryotic cells in their basic structure and organization. This difference between eukaryotic and prokaryotic organisms is the great divide in biology. At the cellular level, differences among animals, plants, fungi, and protists are almost negligible compared with the differences between these groups on the one hand and prokaryotes (bacteria) on the other.

# Containing the cell: Cell membrane

A cell is bound by a membrane, the *cell membrane*, also called the *plasma membrane* or the *plasmalemma*. The cell membrane of all eukaryotes is made of *phospholipid* molecules. The molecules are made by cells, a process that requires energy. The molecules assemble spontaneously (without input of energy) into the membrane, obeying the forces of *polarity*. Turn to Chapter 16 for a discussion of polarity and how the membrane takes its distinctive form, often called the *phospholipid bilayer.*

**REMEMBER**

Don't confuse *cell membranes* with *cell walls.* Every cell has a membrane, and a cell membrane is a fundamental characteristic of a cell. Some cells also have cell walls outside and separate from the cell membrane. No animal cells have cell walls, but some plant cells and fungal cells have them. They're dissimilar to cell membranes in structure and function.

## Permeating the membrane: The fluid-mosaic model

The phospholipid bilayer is embedded with structures of many different kinds. Though the bilayer itself is essentially similar in all cells, the embedded structures are as various and specialized as the cells themselves. Some identify the cell to other cells (very important in immune system functioning); some control the movement of certain substances in or out of the cell across the membrane. Figure 3-2 is a diagrammatic representation of the phospholipid bilayer and embedded structures. This model of the cell membrane is called the *fluid-mosaic model.* "Fluid" describes the ability of molecules in the bilayer to move; "mosaic" pertains to the embedded structures.

The chemical properties of the phospholipid bilayer and the embedded structures contribute to a very important feature of the membrane: It's able to control which substances pass through it and which do not. This means the membrane is *semipermeable.*

## Crossing the membrane passively

Some substances, mainly small molecules and ions, cross the membrane by a *passive transport* mechanism, meaning they more or less flow unimpeded across the bilayer, driven by the forces of "ordinary" chemistry, such as concentration gradients, random molecular movement, and polarity. Here are some ways that substances cross a membrane passively:

**FIGURE 3-2:** The fluid-mosaic model of the cell membrane.

>> **Diffusion:** A substance moves spontaneously down a *concentration gradient* (from an area where it's highly concentrated to an area where it's less concentrated). If you drop a teaspoon of salt into a jar of water, the dissolved sodium and chloride ions will, in time, *diffuse* (spread themselves evenly through the water). You can measure the time in seconds if you stir the solution or in days if you keep the solution perfectly still at room temperature. (To find out why, see Chapter 16.) Cellular and extracellular fluids are constantly being "stirred" and are at temperatures between 95 and 100 degrees Fahrenheit. Molecules to which the cell membrane is permeable (such as oxygen and carbon dioxide) may diffuse into or out of the cell, constantly attempting to reach equilibrium.

The cell membrane is generally not permeable to ions and larger molecules like glucose. They must enter (or leave) the cell through a transport protein via *facilitated diffusion.* This still doesn't require energy because the molecules are moving down their concentration gradient — they just need a door to open for them.

>> **Osmosis:** The diffusion of water molecules across a selectively permeable membrane gets a special name: *osmosis.* As with diffusion, a concentration gradient drives the mechanism. The pressure at which the movement of water across a membrane stops (that is, when the concentration of the solutions on either side of the membrane is equal) is termed the *osmotic pressure* of the system.

>> **Filtration:** This form of passive transport occurs during capillary exchange. (*Capillaries* are the smallest blood vessels — they bridge arterioles and venules; see Chapter 9). Capillaries are only one cell layer thick, and the

capillary wall acts as a filter, controlling the entrance and exit of small molecules. Small molecules dissolved in tissue fluid, such as carbon dioxide and water, are pushed through the capillary wall, sliding between the cells and into the blood, while substances dissolved in the blood, such as glucose and oxygen, do the same in the opposite direction. The pulsating force of blood flow provides a steady force to drive this movement.

The blood pressure in the capillaries is highest at the arterial end and lowest at the venous end. At the arterial end, blood pressure pushes substances through the capillary wall and into the tissue fluid. At the venous end, lower blood pressure (thus higher net osmotic pressure) pulls water from the extracellular fluid (and anything dissolved in it) into the capillary.

**REMEMBER**

Does passive transport contradict the idea that the cell controls what comes in and out through the membrane? No. The substances that move by passive mechanisms are "ordinary" small molecules and ions that are always present in abundance within and between every cell and kept within a physiologically healthy concentration range by the forces of homeostasis, the first line of defense against physiological abnormality. If at any time the physiological levels get too high or too low, the cell has pumps that can counteract the passive transport.

## Crossing the membrane actively

*Active transport* allows a cell to control which big, active, biological molecules move in and out of the cytoplasm. Active transport is a fundamental characteristic of living cells (whereas you can set up a system for diffusion, as we note earlier, in a jar of water).

Like many matters in cell biology, active transport mechanisms are numerous and widely varied. When there is a molecule outside the cell that it needs, a simple active transport mechanism is used. Cell membranes have embedded proteins for the active transport of a single, specific molecule. They must be activated, or opened, for the molecule to be pumped in or out. This is generally done by a binding site on the same protein, but it can also be triggered by another protein with the membrane.

With very large molecules, another energy-requiring transport method is used. For example, a large protein made within the cell could require far too much space to exit — effectively rupturing the cell. Instead, the protein is packaged in a vesicle whose outer membrane is the same phospholipid bilayer as the cell membrane. During this transport process, called *exocytosis*, the lipids realign, letting the protein out of the cell without ever breaching the seal. This same process can occur in reverse, called *endocytosis*, to bring large molecules in.

# Controlling the cell: Nucleus

As we mention previously, the defining characteristic of a eukaryotic cell is the presence of a nucleus (plural, *nuclei*) that directs the cell's activity. The largest organelle, the nucleus is oval or round and is plainly visible under a microscope. Refer to Figure 3-1, earlier in the chapter, to see the relationship of the nucleus to the cell; Figure 3-7, later in the chapter, shows a closer view of the nucleus's structure.

**TECHNICAL STUFF**

All cells have one nucleus, at least at the beginning of their life cycle. As a cell develops, it may lose its nucleus, as do red blood cells and the keratinocytes of the integument; or the cell may merge with other cells, with the merged cell retaining the nuclei of all the cells, such as the fibers of skeletal muscle. This type of cell is called a *syncytium*.

**REMEMBER**

The nucleus contains one complete (diploid) copy of the organism's *genome* — the DNA that embodies the organism's unique genetic material. Every nucleus of every cell in an organism has its own complete and exact copy of the entire genome. It's bound by a semipermeable membrane called the *nuclear envelope.*

The cells produced from this identical DNA are unimaginably varied in structure, in function, and in the substances they produce (proteins, hormones, and so on). The differentiation of the cell (the structure it takes on) and everything about its products are directed by the nucleus, which controls *gene expression,* the selective activation of individual genes.

# Cytoplasm

Within the cell membrane, between and around the organelles, is a fluid matrix called *cytoplasm* or *cytosol* and an internal scaffolding made up of *microfilaments* and *microtubules* that support the cell, give processes the space they need, and protect the organelles. The organelles are suspended in the cytoplasm.

The cytoplasm is gelatinous in texture because of dissolved proteins. These are the enzymes that break glucose down into *pyruvate molecules* in the first steps of cellular respiration (see Chapter 2). Other dissolved substances are fatty acids and amino acids. Waste products of respiration and protein construction are first ejected into the cytoplasm and then enclosed by vacuoles and expelled from the cell.

**TECHNICAL STUFF**

Organelles — including the nucleus, the mitochondria, the endoplasmic reticulum, and the Golgi body — contain a fluid with a particular composition, similar to the cytosol and to one another but each suited to the particular organelle's needs.

## Internal membranes

The plasma membrane isn't the only membrane in a cell. Phospholipid bilayer membranes (without the "mosaic" of embedded structures) are present all through the cell, encapsulating each organelle and floating around, waiting to be useful. The network of membranes is sometimes called the *endomembrane system.* When an organelle makes a substance that must be expelled from the cell, a piece of bilayer moves in and encapsulates (surrounds) the material for exocytosis.

## Powering the cell: Mitochondria

A *mitochondrion* (plural, *mitochondria*) is an organelle that transforms energy into a form that can be used to fuel the cell's metabolism and functions. It's often called the cell's "powerhouse." We discuss the role of the mitochondrion in cellular respiration in Chapter 2.

The number of mitochondria in a cell depends on the cell's function. Cells whose function requires only a little energy, like nervous cells, have relatively few mitochondria; muscle cells may contain several thousand individual mitochondria because of their function in using energy to do "work." A mitochondrion can divide, like a cell, to produce more mitochondria, and it can grow, move, and combine with other mitochondria, all to support the cell's need for energy.

Mitochondria are very small, usually rod-shaped organelles (see Figure 3-3). A mitochondrion has an outer membrane that covers and contains it. The fluid inside the mitochondrion, called the *mitchondrial matrix,* is filled with water and enzymes that catalyze the oxidation of glucose to ATP. A highly convoluted (folded) inner membrane sits within the matrix, increasing the surface area for the chemical reactions.

Unique among organelles, the mitochondrion contains a small amount of DNA in a separate chromosome. This DNA behaves separately and independently from the chromosomes in the nucleus. It duplicates and divides to give birth to new mitochondria within the cell, a separate event from mitosis.

**REMEMBER**

The mitochondrion's two membranes aren't the same as the bilayer membrane of the nucleus.

**FIGURE 3-3:**
A mighty
mitochondrion.

Inner membrane

Matrix

Mitochondrial DNA

Outer membrane

© John Wiley & Sons, Inc.

# The protein factory

The process of protein construction is a truly elegant system, as you see in the "Synthesizing protein" section later in the chapter. Here we look at the structures of the protein-construction system: the organelles and other intracellular structures and their relationships with one another.

The process of protein construction begins in the nucleus. In response to many different kinds of signals, certain genes become active, setting off the production of a specific protein molecule (gene expression). Think of the nucleus as a factory's administrative department.

The *endoplasmic reticulum* (or ER; literally, "within-cell network") is a chain of membrane-bound canals and cavities that run in a convoluted path, connecting the cell membrane with the nuclear envelope. The ER brings all the components required for protein synthesis together. The *ribosomes*, another organelle involved in protein synthesis, adhere to the outer surface of some parts of the membrane, sticking out into the cytoplasm. These areas are called *rough ER* in contrast to *smooth ER*, where no ribosomes adhere. Think of the ER as the factory's logistics function.

The ribosome is the site of protein synthesis, where the binding reactions that build a chain of amino acids are performed. Ribosomes may float in the cytoplasm or attach to the ER. Ribosomes are tiny, even by the standard of organelles, but they're highly energetic, and a typical cell contains thousands of them. Think of the ribosomes as the production machinery.

The *Golgi body* forms a part of the cellular endomembrane system. It functions in the storage, modification, and secretion of proteins and lipids. Think of it as the shipping department. The boxes used to ship the product are the vesicles.

## Lysosomes

Old, worn-out cell parts need to be removed from the cells; if they aren't, they can become sources of toxins or severe energy drains. *Lysosomes* are organelles that do the dirty work of *autodigestion.* Lysosomal enzymes destroy another part of the cell, say an old mitochondrion, through a digestive process. Molecules that can be recovered from the mitochondrion are recycled in that cell or in another cell. Waste products are excreted from the cell in a membrane-bound vacuole.

# Building Blocks That Build You

Though the processes of life may appear to be miraculous, biology always follows the laws of chemistry and physics. Biochemical processes are much more varied and more complex than other types of chemistry, and they happen among molecules that exist only in living cells. Molecules many thousands of times larger than water or carbon dioxide are constructed in cells and react together in seemingly miraculous ways. This section talks about these large molecules, called *macromolecules,* and their complex interactions.

## Joining together: The structure of macromolecules

The four categories of these macromolecules, often called the *biomolecules of life,* include polysaccharides (carbohydrates), lipids, proteins, and nucleic acids. All are made mainly of carbon, with varying proportions of oxygen, hydrogen, nitrogen, and phosphorus. Many incorporate other elements, such as magnesium, sulfur, or copper.

*Macromolecules,* as the term suggests, are huge. Like a lot of huge things, they're made up of smaller things, as a class is made up of students. A student is a subunit of the class, almost identical in many important ways to the other students but unique in other important ways. Macromolecules are made up of molecular subunits called, generically, *monomers* ("one piece"). Each type of macromolecule has its own kind of monomer. Macromolecules are, therefore, *polymers* ("many pieces").

**REMEMBER**

In popular usage, *polymer* suggests plastics. Well, plastics are polymers, but not all polymers are plastics. The term refers to molecules that are made up of repeating subunits — its monomers. The chemical behavior of polymers is different from that of their constituent monomers.

The chemistry of macromolecules is like an infinite Lego set. Any block (monomer) can connect to another block if the shape of their connectors match. With enough blocks, some special connectors, and the energy to do so, you can eventually create a complex structure with bells, whistles, and wheels that turn. Then, you can do that 1,000 times more, and then connect all the complex structures together into one very large, very complex, highly functioning structure. (What's that? You don't have the materials or the energy to do that, and anyway, you wouldn't know how to make a structure fit for the Lego museum? That's okay. Your cells build much more complex things all day every day. All you need to do is keep supplying them with fuel.)

## Polysaccharides

The simple carbohydrate molecule glucose is the main energy molecule in physiology. A common *polysaccharide* (polymer of carbohydrate monomers) is glycogen, which is made by linking numerous glucose molecule together to function as fuel storage.

Polysaccharides are also used for cell structures. Carbohydrate chains can be found attached to proteins embedded in the cell membrane for both recognition by and attachment to other cells.

## Lipids

*Lipids* are polymers of glycerol and fatty acids (three chains of them) and are insoluble in water. The most common lipids are fats, which are an incredibly efficient energy source (which is unfortunate for us but explains the body's propensity for storing it!). Cholesterol is also an important lipid that is used to manufacture steroid hormones (see Chapter 8) and can be found in cell membranes providing stabilization.

The phospholipids I discuss earlier also belong in this category. They replace one of the fatty acid chains with a phosphate group — giving them a hydrophilic head that interacts with water (see Figure 3-2).

# Proteins

Proteins are polymers of *amino acids*. The amino acid monomers are arranged in a linear chain, called a *polypeptide*, and may be folded and refolded into a globular form. Structural proteins comprise about 75 percent of your body's material. The integument, the muscles, the joints, and the other kinds of connective tissue are made mostly of structural proteins like collagen, keratin, actin, and myosin. In addition, the *enzymes* that catalyze all the complex chemical reactions of life are also proteins.

## Amino acids

Twenty different amino acids exist in nature. Amino acids themselves are, by the standards of nonliving chemistry, huge and complex. A typical protein comprises hundreds of amino acid monomers that must be attached in exactly the right order for the protein to function properly.

The binding proclivities among all the amino acids result in the structural precision that makes proteins functional for the exacting processes of biology. That is, the order of amino acids determines how the protein twists and folds. Every protein depends on its unique, three-dimensional structure to perform its function. A misfolded protein won't work and in some cases can result in disease.

## Enzymes

Enzymes are protein molecules that *catalyze* the chemical reactions of life. Enzymes can only speed up a reaction that is otherwise chemically possible. How effective are enzymes in speeding up reactions? Well, a reaction that may take a century or more to happen spontaneously happens in a fraction of a second with the right enzyme. And better yet, they are not "used up" in the process. Enzymes are involved in every physiological process, and each enzyme is extremely specific to one or a very few individual reactions. Your body has tens of thousands of different enzymes.

**REMEMBER**

Any time an enzyme is discovered, it is named, usually with some physiology shorthand for its function, frequently with the suffix *–ase.* An enzyme named *pyruvate dehydrogenase* removes hydrogen atoms from pyruvate molecules in the processes of cellular respiration (see Chapter 2).

**TECHNICAL STUFF**

The malfunctioning of a single enzyme, which can be caused by a single nucleotide being out of order, is responsible for some very nasty, sometimes fatal diseases. *Phenylketonuria* (PKU) is an inherited metabolic disease caused by faulty *phenylalanine hydroxylase.* The inability to properly metabolize the amino acid phenylalanine, found in many foods, results in mental retardation, organ damage, unusual posture, and even death, unless the afflicted person can limit the ingestion of foods containing phenylalanine, which is a very difficult thing to do.

# Nucleic acids

The nucleic acids *deoxyribonucleic acid* (DNA) and *ribonucleic acid* (RNA) are poly-mers made of monomers called *nucleotides* and arranged in a chain, one after another . . . and another, and another. DNA molecules are thousands of nucleotides long (see Figure 3-4). The functioning of genes is inseparable from the chemical structure of the nucleic acid monomers.

Key:

G  = Guanine

C  = Cytosine

A  = Adenine

T  = Thymine

S = Deoxyribose sugar

P = Phosphate group

Hydrogen bonds

Strand 1

Strand 2

**FIGURE 3-4:** DNA is made up of thousands of nucleotides.

© John Wiley & Sons, Inc.

A nucleotide is made up of a sugar molecule and a phosphate group attached to a nitrogenous base. The sugar molecule is either deoxyribose (in DNA) or ribose (in RNA). The nitrogenous base is one of four:

>> Cytosine (C), guanine (G), thymine (T), or adenine (A) in DNA

>> Cytosine (C), guanine (G), uracil (U), or adenine (A) in RNA

The bases connect with each other in specific pairs. (Refer to the section "Gene structure" later in the chapter for a discussion of the biological significance of this complementary pairing.) The *complementary pairs* are

>> C with G

>> A with T in DNA

>> A with U in RNA

The structural similarities and differences between DNA and RNA allow them to work together to produce proteins within cells. The DNA molecule remains stable in the nucleus during normal cell functioning, protected from damage by the nuclear envelope. An RNA molecule is built on demand to transmit a gene's coded instructions for building proteins, and then it disintegrates. Some of its nucleotide subunits remain intact and are recycled into new RNA molecules.

# Genes and Genetic Material

Your anatomical structures are specified in detail, and all your physiological processes are controlled by your very own unique set of *genes.* Unless you're an identical twin, this particular set of genes, called your *genome,* is yours alone, created at the moment your mother's ovum and your father's sperm fused. The genome itself (all your genes) is incorporated in the DNA in the nucleus of each and every one of your cells.

TECHNICAL
STUFF

The genome consists of many thousands of genes. Early research into the number of individual genes in the human genome has produced varying estimates, from about 23,000 to around 75,000, with estimates toward the lower end of the range predominating. Within any given cell, only a very few of these genes is ever *expressed* — that is, activated and used in protein formation.

# Traiting you right

Your genes are responsible for your *traits.* If your genes specify that you'll grow to 6 feet tall (a trait), they cause bone and tissue to grow until your body reaches that height, and they maintain it thereafter (assuming a favorable environment), until the cells the genes work through age and die. If you have the genes for brown eyes (a trait), your genes direct the production of pigments that color the eyes. And if your genes include those for abundant production of low-density cholesterol, you have a tendency toward atherosclerosis, another trait. In an environment where the available nutritional resources include abundant red meat, this trait could give you trouble. Otherwise, this trait is harmless.

**TECHNICAL STUFF**

Calling a trait "harmless" may be technically incorrect. Assuming that any trait has some survival-enhancing purpose, at least under some environmental conditions, is probably more correct, even if you don't know what the purpose is.

# Gene structure

The physical structure of the DNA molecule, called the *DNA double helix,* is key to the functioning of genes (refer to Figure 3-4). The DNA double helix can be compared with a ladder or a zipper. In function, it's more like a system of code.

## Code "you"

The nucleotides are the symbols in the genetic code. In modern English, a written word is a code made of certain subunits (letters) set down in a specific order. For example, *this, than, then,* and *ten* are different words with different meanings and functions in expressing a thought. A *gene* is made of certain nucleotides in a certain order. A gene may be a few or many nucleotides in length (imagine a single word 20 pages long), but a given gene is always exactly the same nucleotides in exactly the same order along one strand of DNA. The order of the nucleotides is absolutely crucial: "ACTTAGGCT" is not the same as "ACTAAGGCT." According to the prevailing theory, each gene specifies the construction of one protein molecule. This is called the *one gene, one protein* model. If the nucleotides are out of order, the protein molecule they make will probably be useless, and the organism's functioning will likely be impaired, a little or a lot.

**REMEMBER**

Every model used to explain cell biology, including the *one gene, one protein* model, is subject to change as more information becomes available. But the changes are likely to be minor for understanding basic anatomy and physiology.

### Pairing at the molecular level

Remember nucleotide complementary pairs? (If not, see the section "Nucleic acids and nucleotides" earlier in the chapter.) So if a nucleotide is in place on one strand of DNA, and each type of nucleotide binds with only its complementary partner (A with T, C with G, and so on), what do you think is on the other DNA strand? Right! The other nucleotide of the complementary pair. Wherever there's a G on one strand, there's a C on the other, and the pair is attached in the middle. And if the two strands become separated (which they do), what do you think will happen? The nucleotides on each strand will attract and hold other molecules of their complementary partners, thus creating two new double-strands identical to the original double-strand. This is how DNA replication occurs.

## Synthesizing protein

A gene that's active sends messages to its own cell or to other cells, ordering them to produce molecules of its particular protein, the only one it's capable of making. The message is sent from DNA through the intermediary of *mRNA* (messenger RNA), which places an order at the protein factory of the cell and stays around for a while to supervise production. The first part of the process, where DNA "writes the order" in the form of a sequence of nucleotides on mRNA, takes place in the nucleus and is called *transcription* ("writing across"). The next part of the process, where RNA places the order at the factory, is called *translation* ("carrying across"). The last part, where the amino acid monomers are sorted out and assembled into the polypeptide, starts in the ribosome. Figure 3-5 shows this process. The finishing touches are put on the protein molecule in the endoplasmic reticulum and the Golgi body. The process from transcription to the last finishing touch on the protein molecule is called *gene expression*.

**REMEMBER**

Keep in mind the relationship between the *genome* and *gene expression*. Your entire genome is contained in identical DNA molecules in the nucleus of every one of your cells; it remains unchanged through your lifetime. Any given gene may be expressed in only a few cells, or only occasionally in a few cells, or only under certain physiological conditions, or possibly never. The totality of gene expression changes every second throughout your lifetime, as quickly as nerves transmit impulses and cells react.

**REMEMBER**

The terms *gene expression, protein synthesis,* and *transcription-and-translation* are essentially the same in meaning, but they're used in different contexts in cell biology and physiology.

**FIGURE 3-5:**
The process of protein synthesis.

# The Cell Cycle

The life cycle of an individual cell is called the *cell cycle.* The moment of *cell cleavage,* when a cell membrane grows across the "equator" of a dividing cell, is considered to be the end of the cycle for the mother cell and the beginning of the cycle for each of the daughter cells.

Typically, but by no means universally, *interphase* is the longest period of the cell cycle. Interphase comes to an end when the cell divides in the process of *mitosis.* We discuss both these periods of the cell cycle in the following sections.

## Cells that divide, cells that don't

All cells arise from the division of another cell, but not all cells go on to divide again:

>> **Zygote:** This is the diploid cell that comes into existence when the sex cells (ovum and sperm, both haploid) fuse at conception. Almost immediately, the zygote divides into two somatic cells.

>> **Somatic cells:** These include all the cells of the body except the sex cells — in other words, all the diploid cells of the body. Somatic cells may be *relatively differentiated* (somewhat specialized), *terminally differentiated* (they never divide again), or stem cells.

>> **Stem cells:** These are special kinds of rather "generic" somatic cells that divide to produce one new stem cell and one new somatic cell that goes on to differentiate into a particular type of cell in a particular type of tissue. The *embryo* (organism in the very early stages of development) has very special stem cells, called *pluripotent* ("many powers") stem cells, which have the ability to give rise to just about any kind of cell an organism needs, given the right chemical environment. When an organism has developed beyond the embryo stage, embryonic stem cells disappear, and other types of stem cells, called *multipotent* or *adult* stem cells, arise in particular tissue types and specialize in producing new cells for that tissue. (Turn to Chapter 9 for a description of how stem cells in the bone marrow give rise to many types of blood cells.)

>> **Sex cells (gametes):** These form when specialized somatic cells in the reproductive system divide by a process called *meiosis.* Meiosis is the only cellular process in the human life cycle that produces *haploid* cells. See Chapter 14 for more details on sex cells and the processes of meiosis.

# UNCONTROLLED CELL GROWTH

Our cells have genes that control the *cell cycle* — that is, how often cells undergo mitosis. Mutations in these genes allow cells to divide unabated. This uncontrolled cell growth is the very definition of *cancer*. As a result, any tissue in your body has the potential to become cancerous. That most common cancers (worldwide) are lung, breast, colorectal, and prostate.

Cancers are grouped by their origin and the cell type affected. The cells dividing out of control may invade the surrounding tissue, creating a *tumor*. They may also be knocked loose and take up residence elsewhere, called *metastasis*. This complicates treatment as you may find a tumor in the prostate gland, but it's actually cancerous bladder tissue that metastasized there.

Because the cells of each tissue are unique, there is no single cure-all for cancer. Surgery to remove the tumor is often performed if it's accessible. *Radiation* treatment may be used to kill the cancers cells in a tumor but kills nearby healthy cells, too. *Chemotherapy* uses chemicals that target the specific cell type affected. Treatment often involves a combination of all three.

Researchers the world over are trying to find successful, specific treatments that leave the healthy cells unharmed. Unfortunately, we have to battle cancer one tissue type at a time.

Table 3-2 summarizes how different types of cells behave when it comes time to divide.

**TABLE 3-2** **Dividing Behavior of Different Cell Types**

| Cell Type | Arise From | Divide? | Give Rise To |
|---|---|---|---|
| Zygote | Fusion of two sex cells | Y | Two somatic cells |
| Somatic cell | Somatic cell or stem cell | Y or N* | Somatic cells; sex cells** |
| Stem cell | Stem cell | Y | One specialized somatic cell and one stem cell |
| Sex cell | Somatic cell | N | NA |

*Some somatic cells go on to terminal differentiation and never divide again.*

**Sex cells arise from meiosis of certain somatic cells. They are haploid cells and never divide again.*

# Interphase

*Interphase* begins when the cell membrane fully encloses the new cell and lasts until the beginning of mitosis or meiosis. The duration of interphase may be anywhere from minutes to decades. Generally speaking (there are always exceptions in cell biology), cells do most of their differentiating and most of their routine metabolizing during interphase. Stem cells grow in size and duplicate organelles during interphase, in preparation for mitosis (more on mitosis in a minute). Some other cells enter mitosis after an extended period of steady-state metabolism. Sometimes, a cell remains in interphase, carrying out its physiological function for years and years until it dies.

# DNA replication

*DNA replication* is an early event in cell division, occurring during interphase, just prior to the beginning of mitosis or meiosis but within the protected space in the nuclear envelope. Maintaining the integrity of the DNA code is absolutely vital.

During DNA replication, the double helix must untwist and "unzip" so that the two strands of DNA are split apart. As shown in Figure 3-6, each strand becomes a template for building the new complementary strand. This process occurs a little at a time along a strand of DNA. The entire DNA strand doesn't unravel and split apart all at once. When the top part of the helix is open, the original DNA strand looks like a Y. This partly open/partly closed area where replication is happening is the *replication fork.*

**TECHNICAL STUFF**

In Figure 3-6, the symbols 5' and 3' (read *five prime* and *three prime*) indicate the direction in which DNA replication is occurring. The template strand is read in the 3'-to-5' direction. The bases that are complementary to the template strand are added in the 5'-to-3' direction.

# Mitosis

A cell enters a process of *mitosis* (division) in response to signals from the nucleus. As shown in Figure 3-7, mitosis is a multistage process, proceeding in the following stages:

1. **Prophase: The nuclear envelope is dismantled, and the duplicated DNA, in the form of *chromatin*, thickens and coils into *chromosomes*.** Each duplicated chromosome is composed of two identical strands of DNA referred to as *chromatids*. The chromatids are held together by a protein mass called the *centromere*. (**Note:** When the chromatids separate, each is considered a new chromosome.) Cellular structures called *centrioles* and *spindle fibers* form and move to the poles of the cell.

**FIGURE 3-6:** The DNA replication process.

© John Wiley & Sons, Inc.

**2.** **Metaphase: The chromosomes line up to form a perfect row at the center of the cell. The spindle fibers, attached to the centrioles at one end, attach to the chromosomes.** At this point, 92 chromatids are in a double set of 46 chromosomes.

3. **Anaphase: The centromere is split by enzymatic activity, and the chromatids are pulled apart by the spindle fibers toward one of the centrioles: 46 to one, 46 to the other.** Following this movement, the chromosomes are referred to as *daughter chromosomes,* and the set at one pole is identical to the set at the opposite pole. But the cell isn't quite ready to divide yet.

4. **Telophase: A fresh nuclear membrane is reassembled around each set of chromosomes.** The spindles dissolve, which frees the daughter chromosomes.

At this point, when each of the two identical nuclei is at one pole of the cell, mitosis is technically over. However, the cell's cytoplasm still has to actually split apart into two masses, a process called *cytokinesis.* The center of the mother cell indents and squeezes the cell membrane across the cytoplasm until two separate cells are formed. The two daughter cells are then in interphase and go on to differentiation or not, depending on the instructions to the cell from the genome.

**Late Interphase**

Chromatin — Nuclear pore
Nucleus
Nucleolus
Centriole

Prophase **Mitosis** Metaphase
Spindles

Fragments of
nuclear envelope
Centromere
Chromatids

Chromosomes are
aligned at equator

Telophase Anaphase

**Early Interphase**

Daughter cells

**FIGURE 3-7:**
The stages of
mitosis:
prophase,
metaphase,
anaphase, and
telophase.

*Illustration by Kathryn Born, MA*

# Organizing Cells into Tissues

A *tissue* is an assemblage of cells, not necessarily identical but from the same origin, that together carry out a specific function. As we discuss in Chapter 1, tissue is the second level of organization in organisms, above (larger than) the cell level and below (smaller than) the organ level.

Like just about everything else in anatomy, tissues are many and various, and they're grouped into a reasonable number of "types" to make talking about them and understanding them a little simpler. The tissues of the animal body are grouped into four types: *connective tissue, epithelial tissue, muscle tissue,* and *nervous tissue.* All body tissues are classified into one of these groups.

## Connecting with connective tissue

Connective tissues connect, support, and bind body structures together and are the most abundant tissue by weight. Generally, connective tissue is made up of cells that are spaced far apart within a gel-like, semisolid, solid, or fluid matrix. (A *matrix* is a material that surrounds and supports cells. In a chocolate chip cookie, the dough is the matrix for the chocolate chips.)

Connective tissue has many functions, and thus many forms; it is the most varied of all the tissue groupings. In some parts of the body, such as the bones, connective tissue supports the weight of other structures, which may or may not be directly connected to it. Other connective tissue, like adipose tissue (fat pads), cushions other structures from impact. You encounter lots of connective tissue in the chapters to come because every organ system has some kind of connective tissue.

We discuss the specialized connective tissues *bone* and *cartilage* in some detail in Chapter 5, and we discuss the important connective tissue *blood* in detail in Chapter 9. (What? Blood is a tissue? A connective tissue? Yes, and you'll see why.)

The other types of connective tissue, or *proper connective tissues,* are all means of connections. Just as we have different kinds of tapes and glues, we have different varieties of connective tissue. They contain varying proportions of fibrous proteins of two types: collagenous and elastic. *Collagenous fibers* are made of collagen, a bulky protein, and serve to provide structure. *Elastic fibers* are made of elastin, a thin protein, and serve to provide stretch.

The following are the proper connective tissues found in the human body:

>> **Areolar (a type of loose connective tissue):** This tissue surrounds and separates structures in every part of the body. It forms a thin membrane with ample space for blood vessels to pass through. Both collagenous and elastic fibers are prevalent in its gel-like matrix.

>> **Dense regular connective tissue:** This tissue is characterized by its tightly packed collagenous fibers arranged parallel to each other, making it very strong. There are very few cells present and there is very little blood flow. Tendons and ligaments are made of dense regular tissue (see Chapter 5).

>> **Dense irregular connective tissue:** Very similar to dense regular, the fibers in this tissue are disorganized, leaving more space for blood flow. The dermis is a typical tissue of this type (see Chapter 4).

>> **Adipose tissue (a type of loose connective tissue):** Composed of fat cells, adipose tissue provides fuel storage and insulation as well as support and protection to its underlying structures.

>> **Reticular tissue (a type of loose connective tissue):** This type of tissue uses a thinner variety of collagenous fibers to form a net. It creates the framework of such organs as the spleen, the lymph nodes, and the liver.

>> **Elastic connective tissue:** Collagenous fibers are bundled in parallel with bands of elastic fibers sandwiched between (like lasagna). This makes elastic tissue strong but stretchy — perfectly suited for the walls of hollow organs and arteries.

## Continuing with epithelial tissue

*Epithelial tissue* forms the epidermis of the *integument* (the skin and its accessory structures; see Chapter 4), covers all your internal organs, and forms the lining of the internal surfaces of the blood vessels and hollow organs.

Epithelial tissues create coverings and linings; they're always bordered by "empty space" on one side. The other side is a *basement membrane* that allows resources to diffuse up into the tissue from the connective tissue below (epithelial tissues have no blood flow).

Epithelial tissue comes in eight types that are defined by the way epithelial cells are combined and shaped (see Figure 3-8).

**FIGURE 3-8:**
The cellular composition of epithelial tissue.

Illustration by Kathryn Born, MA

>> **Simple squamous epithelium:** A single layer of flat cells, this tissue functions in rapid diffusion and filtration. The lining of the alveoli (small sacs) of the lung is a typical tissue of this type.

>> **Simple cuboidal epithelium:** A single layer of cuboidal cells, this tissue functions in absorption and secretion. This tissue is typically found in glands. The cuboidal cells have the capacity to produce and modify the glandular product (for example, sweat, oil, or milk).

>> **Simple columnar epithelium:** A single layer of cells that are elongated in one dimension (like a column). Like simple cuboidal epithelium, this tissue functions in secretion and absorption. This type of tissue is primarily found lining portions of the digestive tract.

The cells may also be *ciliated,* possessing a type of organelle called *cilia* — hairlike structures that act to move substances along in waves. Ciliated simple columnar epithelium can be found lining the uterine tube.

>> **Pseudostratified columnar epithelium:** A single layer of columnar cells. Note that the prefix *pseudo* means "false." The tissue appears stratified, or layered, because the cells' nuclei don't line up in a row, as they do in simple columnar epithelium. Other than that, they're the same, and have similar functions of absorption and secretion. This type of tissue lines ducts in testicular structures.

More commonly, this tissue type is ciliated. It is present in the linings of the respiratory tract, functioning in a more or less identical way to the simple columnar ciliated epithelium.

- >> **Stratified squamous epithelium:** This tissue consists of several layers of cells: squamous epithelial cells on the outside with deeper layers of cuboidal or columnar epithelial cells. It's found in areas where the outer layer is subject to wear and needs to be replaced continuously. The epidermis of the skin is an example of a specific type called *keratinized stratified squamous epithelial tissue.*

- >> **Stratified cubiodal epithelium:** Several layers of cuboidal cells that line the ducts associated with the sweat, mammary, and salivary glands.

- >> **Stratified columnar epithelium:** Several layers of elongated cells mainly act as protection in such structures as the conjunctiva of the eye and the pharynx.

- >> **Transitional epithelium:** The cells of this tissue can change (or *transition*) from cuboidal when relaxed to squamous when stretched, as needed by the tissue. This tissue is found in the lining of the bladder, where having a little room to stretch is sometimes handy. See a diagram of transitional epithelium in Chapter 12.

## Mixing it up with muscle tissue

Muscle tissue comes in three types: skeletal muscle, smooth muscle, and cardiac muscle. We discuss the similarities and differences in cellular composition in these three types of tissue in Chapter 6. We also discuss in Chapter 6 the anatomy and physiology of the large organ system called the *muscular system*, of which skeletal muscle tissue is a major component. In Chapter 9, we discuss the function of cardiac muscle in the context of the cardiovascular system as well as the role of smooth muscle in blood circulation. We also cover smooth muscle's role in the digestive system in Chapter 11.

## Getting nervous about nervous tissue?

Don't be. Nervous tissue is relatively simple in one way: Your body has only one type of nervous tissue, and it's made mostly of only one type of cell, the *neuron.* You can get lots more information about the nervous system in Chapter 7, if you get the *impulse* to find out more.

# 2

# Sizing Up the Structural Layers

**IN THIS PART . . .**

Home in on the major anatomical structures of the integumentary, skeletal, and muscular systems.

Suss out the specialized functions of the skin.

Consider the structure of bones and how that structure relates to their function.

Find out how the skeleton moves.

Look at the physiology of muscle contraction.

IN THIS CHAPTER

» Seeing what the integument does

» Explaining the integument's structure

» Taking a closer look at hair, nails, and glands

» Understanding what your skin does for you

» Checking out the integument's pathophysiology

# Chapter 4

# Getting the Skinny on Skin, Hair, and Nails

The skin is the largest organ of the human body. In a human adult, the skin covers an area of about 20 square feet (almost 2 square meters), or about the size of a small blanket — a soft, pliable, strong, waterproof, and self-repairing blanket. It accounts for about 5 percent to 7 percent of the body's weight.

The skin and its appendages (hair, nails, and so on) form your outer covering, or *integument.* Though it is one of the most beautiful parts of you, it's also a complex organ system with numerous types of tissue and many specialized structures. The average square inch (about 6.5 square centimeters) of skin holds 650 sweat glands, 20 blood vessels, more than 1,000 hair follicles, half a million melanocytes (pigment cells), and more than 1,000 nerve endings. Refer to the "Skin (Cross Section)" color plate in the center of the book to see a detailed cross section of the skin.

Skin is made in layers, and layers within layers. New cells begin life in the lower levels and are gradually pushed to the surface to replace the old, dead ones. By the time the new cells reach the top, they've become hard and flat, like roof shingles. Eventually, they pop off like shingles blown from a roof in a strong wind.

Amazingly, every minute, 30,000 to 40,000 dead skin cells fall from your body. In approximately a month's time, all the cells on the top layer have blown off and been replaced.

In this chapter, we show you what your skin is made of, and we explain how your hair and nails form. We also look at the functions of your skin and what diseases and conditions can take a toll on your integument.

# Functions of the Integument

The entity known as *you* is bounded by your integument. The skin mediates much of the interaction between *you* and *not-you* — the environment. Your integument identifies *you* to other humans, a very important function for members of the hypersocial human species. Here's a look at the integument's other important functions:

>> **Protection:** Skin protects the rest of the body by keeping out many threats from the environment, such as pathogens, damaging solar radiation, and nasty substances everywhere.

>> **Thermoregulation:** The skin and its appendages in particular support *thermoregulation* (maintaining a consistent body temperature) in several ways. See the "Controlling your internal temperature" section later in this chapter.

>> **Water balance:** The skin's outer layers are more or less impermeable to water, keeping water and salts at an optimum level inside the body and preventing excess fluid loss. A small amount of excess water and some cellular waste are eliminated through the skin.

>> **Incoming messages:** Many types of sensory organs are embedded in your skin, including receptors for heat and cold, pressure, vibration, and pain. See the "Your skin is sensational" section later in the chapter.

>> **Outgoing messages:** The skin and hair are messengers to the outside environment, mainly to other humans. People get information about your state of health by looking at your skin and hair. Your emotional state is signaled by pallor, flushing, blushing, goose bumps, and sweating. The odors of sweat from certain sweat glands signal sexual arousal.

>> **Substance production:** *Sebaceous glands* in the skin, usually associated with a hair follicle, produce a waxy substance called *sebum* for waterproofing. Sweat glands in the skin make sweat. In fact, your skin has several different types of glands, and each makes a specific type of sweat. See the "Nothing's bland about glands" section later in the chapter.

Skin cells produce *keratin,* a fibrous protein that's an important structural and functional component of skin and is, essentially, the only component of hair and nails. See the "Caring about Keratins" sidebar later in the chapter.

**TECHNICAL STUFF**

The root of integument is *integr-,* meaning "whole" as in *integer* (a whole number), *integrate* (to bring different parts together into one), *integral* (inseparable from the whole), and *integrity* (always a positive term, but more difficult to define).

# Structure of the Integument

The integument wraps around the musculoskeletal systems, taking the shape of your bones and muscles and adding its own shape-forming structures. Although your skin feels tight, it's really loosely attached to the layer of muscles below. In spots where muscles don't exist, such as on your knuckles, the skin is attached directly to the bone.

Without losing sight of the reality, it's sometimes helpful to imagine that you can unzip and remove your skin, spread the living skin out on a table, and look at it. What would you see, feel, and smell?

One of the most obvious features you'd notice is that the skin, itself a thin layer, is made up of several layers. The layering is visible to the unaided eye because each layer is different from the others and the transitions between layers appear to be relatively abrupt. The superficial (or outermost) layer is the epidermis followed by the dermis and then the hypodermis. We look at each of these layers in the following sections.

**REMEMBER**

The epidermis is on top of the dermis. The prefix epi- means "on top of." Many anatomical structures are on top of other structures, so you see this prefix in many chapters throughout this book. Other related terms you may see are endo (within), ecto (outside), and hypo (beneath or lower).

**REMEMBER**

In this discussion, *up* and *above* mean "toward the surface of the body," and *down* and *below* mean "toward the center of the body." These terms *don't* mean "toward the head" and "toward the feet," respectively. Anatomists use the terms *superficial* and *deep* to mean the same thing.

## Touching the epidermis

The most familiar aspect of the integument is the *epidermis;* it's the part you see when you look in the mirror or at others. The epidermis feels soft, slightly oily, elastic, resilient, and strong. In some places, the surface contains dense, coarse hairs; in other spots, it has a lighter covering of finer hairs; and in a few places, it has no hairs at all. The nails cover the tips of the fingers and toes.

The epidermis is composed of *stratified squamous* tissue and has no direct blood supply. Its nourishment diffuses through the basement membrane from the dermis below. As your skin cells age and get pushed further away from the resource supply, they weaken and eventually die. This gives the epidermis its layered appearance (not visible to the naked eye though). Figure 4-1 illustrates the layers of the epidermis.

Illustration by Kathryn Born, MA

**FIGURE 4-1:** Layers of epidermis.

The healthiest, happiest layer of epidermal cells is along the basement membrane, called the *stratum basale* (or *stratum germinativum*). This layer is the only one that has cells that reproduce. The older cells shift upward, becoming spindle-shaped, forming the *stratum spinosum,* the thickest of the epidermal layers. Next is the *stratum granulosum;* cells in this layer are flattened and the nuclei and organelles start to shrivel. The most superficial layer is the *stratum corneum;* these are hardened, dead cells full of keratin. Thick skin, such as the palm of your hands, has an additional layer between the granulosum and corneum called the *stratum lucidum;* named because the cells appear lucid (or clear) under a microscope, this layer is built up for added protection in areas receiving a lot of wear and tear.

**REMEMBER**

Cells are shed continuously from the top of the epidermis and replaced continuously by cells that are pushed up from the deeper layers. The entire epidermis is replaced approximately every six to eight weeks throughout life.

### The thin, impervious cover

Think of the *stratum corneum* as a sheet of self-repairing fiberglass over the other layers of the epidermis. It's only 25 to 30 cells thick, dense, and relatively hard. All the cells are *keratinized,* as they became filled with this protein (keratin) while they are pushed toward the surface. The keratin and other proteins mechanically stabilize the cell against physical stress. Sebaceous glands in the dermis release *sebum* onto the surface of the stratum corneum for softening and waterproofing. The stratum corneum protects the entire body by making sure that some things stay in and everything else stays out. (See the nearby sidebar "Cosmetics and the stratum corneum.")

One important thing that the stratum corneum doesn't seal out, however, is ultraviolet radiation. This form of energy goes right through the skin's surface and down to the layers below, where it stimulates the production of vitamin D. In high doses, it burns the skin and damages DNA, which can cause cells to become cancerous. Some exposure to UV radiation is necessary for our health, but too much UV radiation can clearly be problematic. Special cells in the stratum basale produce *melanin,* which absorbs harmful UV radiation and transforms the energy into harmless heat. More on this later in this section.

## COSMETICS AND THE STRATUM CORNEUM

The use of cosmetics is far older than recorded history, along with other appearance-enhancers and subgroup identifiers like tattoos and scarifications (burning or etching). Two types of cosmetics commonly used in culture today are moisturizers and exfoliants.

**Moisturizers**

What do moisturizers do? After the water in them evaporates into the air, the lipids they contain remain on the skin's surface, adding another waterproof barrier to prevent the escape of water from the skin. This increases the very scant water in the layer and helps block particles and chemicals from getting through. That's it.

As far as their effects on the skin, all moisturizers are essentially the same, whether they're petroleum jelly, a luxury-brand "serum," a drugstore lotion, or a vitamin E cream from the health food store. Not that they don't contain the ingredients the

*(continued)*

*(continued)*

manufacturers claim; it's just that none of them gets past the stratus corneum. All those age-defying peptides, metallic microparticles, organic extracts, vitamins, rare botanicals, and antioxidants remain within the lipid layer on the skin's surface and go down the drain with the lipid and the trapped dirt when the skin is washed.

**Exfoliants**

Speaking of washing, what about exfoliants? *Exfoliation* means to remove the topmost cells of the stratum corneum by mechanical or chemical means. Your skin sheds these dead cells all the time, but exfoliation hurries up the process for those cells it reaches. Generally, exfoliation neither hurts nor helps any physiological process in the skin: The stratum corneum remains an effective barrier. Some people think that having some slightly younger (but still dead) cells on the skin's surface enhances its appearance.

Shaving is one common form of mechanical exfoliation as cells are removed as the hairs are cut. A mechanical exfoliant cosmetic (such as sugar scrubs) usually contains an abrasive substance embedded in the soap. Different people, and different skins, prefer different abrasives. Chemical exfoliation (such as masks and peels) usually involves the application of an acidic substance that breaks up the fibers holding the cells of the stratum corneum together. But no worries — ranks of keratinized cells are always moving up from below to take their place. The chemicals may also produce a slight irritation that leads to a temporary inflammation reaction, puffing out some shallow wrinkles in the skin (again, the goal being attractiveness).

Does that mean moisturizers and exfoliants are a waste of money? Not necessarily. Keeping the skin moist, lubricated, and clean feels good and is a good thing generally. You may choose one cosmetic over another for a number of reasons: nicer texture, nicer perfume, the absence of perfume or ingredients that irritate the skin. And for some people, the use of an expensive self-care product enhances self-esteem ("Because you're worth it," as the advertising tag line of one brand says). Self-esteem and self-confidence enhance psychological well-being, which may enhance reproductive success and survival.

## The transit zone

The *stratum lucidum*, found only on the palms of hands and soles of the feet (thick skin); the *stratum granulosum*; and the *stratum spinosum* lie in distinct layers below the stratum corneum. Old cells slough off above and new cells push up from below, finally getting up into the stratum corneum. The process takes about 14 to 30 days. Most of the cells that comprise the epidermis are called *keratinocytes*. These cells create structural proteins (like keratin), lipids, and even some antimicrobial molecules. As they are pushed away from the nutrient source (blood vessels in the dermis) they become progressively more *keratinized.* These layers also contain *Langerhans cells,* immune cells that arrest microbial invaders and transport them to the lymph nodes for destruction.

## The cell farm

The *stratum basale,* also called the *stratum germinativum* or *basal layer,* is like a cell farm, constantly producing new cells and pushing them up into the layer above. This layer also contains *melanocytes,* which produce the *melanin* pigment that gives color to your skin, hair, and eyes and protects the skin from the damaging effects of UV radiation in sunlight. Melanin absorbs UV radiation and dissipates more than 99.9 percent of it as heat.

Everybody's stratum basale has about the same number of melanocytes (half a million to a million or more per square inch of skin), but the amount of melanin produced varies depending mainly on genetics (heredity). The environment can also play a role in variety of skin color because exposure to UV radiation stimulates increased production of melanin. Human groups living close to the equator have evolved genes that stimulate melanocytes to produce more melanin as protection from UV radiation. Without the melanin, the radiation can burn the skin, damage DNA, and ultimately cause skin cancer.

# Exploring the dermis

Below the layers of the epidermis and several times thicker is the *dermis.* The dermis itself is made up of two layers: the papillary region and the reticular region.

## CARING ABOUT KERATINS

*Keratins* are fibrous proteins produced by skin cells in mammals, birds, and reptiles. Keratin-containing anatomical structures are hard, tough, and waterproof. The α-keratins, are the main components of hair — including wool, nails, claws, hooves, and horns. (But not antlers, they're derived from bone.) The baleen of filter-feeding whales is made mainly of α-keratins. The other type, the β-keratins, are even harder and are found in human nails, mammalian claws, and porcupine quills; in the shells, scales, and claws of reptiles; and in the feathers, beaks, and claws of birds. People who make educated guesses about the anatomy and physiology of dinosaurs believe that keratins were likely a major component of the claws, horns, and armor plates of these creatures.

## Under the basement

The *papillary region* is made up of the *basement membrane,* which sits just below the epidermis, and *papillae* (finger-like projections) that push into the basement membrane, increasing the area of contact between the dermis and the epidermis. In your palms, fingers, soles, and toes, the papillae projecting into the epidermis form *friction ridges.* (They help your hand or foot to grasp by increasing friction.) The pattern of the friction ridges on a finger is called a *fingerprint.*

REMEMBER

The papillary region is an example of a common anatomical "strategy" for increasing the surface area between two structures. More areas of direct contact lead to more chances for molecules to travel from one side to the other. Think about the difference between leaving a crowded parking lot after a concert with only 2 exits and one with 20 exits. We discuss another prominent example, the intestines, in Chapter 11.

## A manufacturing site

The *reticular region* is chock-full of protein fibers and is a complex and metabolically active layer. Cells and structures of the reticular region manufacture many of the skin's characteristic products: hair and nails, sebum, watery sweat, apocrine sweat. (See the "Accessorizing Your Skin" section later in the chapter.) The region also contains structures that connect the integument to other organ systems: sensory structures to communicate with the nervous system, lymphatic vessels, and a very rich blood supply.

The blood vessels in the dermis provide nourishment and waste removal from its own cells as well as from the stratum basale. The blood vessels in the dermis dilate (become larger) when the body needs to lose heat and constrict to keep heat in. They also dilate and constrict in response to your emotional state, brightening or darkening skin color, thereby functioning as social signaling.

## Getting under your skin: The hypodermis

The *subcutaneous layer* (or *hypodermis*, or *superficial fascia*) is the layer of tissue directly underneath the dermis. It is mainly composed of connective and *adipose* tissue (fatty tissue). Its physiological functions include insulation, the storage of energy, and help in the anchoring of the skin. It contains larger blood vessels, lymphatic vessels, and nerve fibers than those found in the dermis. Its loosely arranged elastin fibers anchor the hypodermis to the muscle below.

The thickness of the subcutaneous layer is determined in some places by the amount of fat deposited into the cells of the *adipose tissue,* which makes up the majority of the subcutaneous layer. Recently, researchers have found that adipose tissue also plays a very active role in the endocrine system (see Chapter 8).

# Accessorizing Your Skin

This section has nothing to do with tattoos or body piercing. It's all about your skin's *accessory structures:* hair, nails, and glands — structures that work with the skin.

## Now hair this

Your body has millions of hair follicles, about the same number as the chimpanzee, humanity's closest evolutionary relative. (Although we don't know who counted all of them!) Like chimps, humans have hairless palms, soles, lips, and nipples. Unlike a chimpanzee, most of your hair is lightweight and fine. The hair on your head is coarser and longer to help hold in body heat. Puberty brings about a surge of sex hormones that stimulate hair growth in the *axillary* (armpit) and pelvic regions and, in males, on the face and neck. Women with hormonal imbalances can develop facial hair, too. Turn to Chapter 8 for more about hormones.

A hair arises in a *hair follicle,* a small tube made up of epidermal cells that extend down into the dermis to take advantage of its rich blood supply. Like the epidermis, cells at the bottom of the hair follicle, called the *papilla,* continually divide to produce new cells that are added to the end of the hair and push the older cells up through the layers of the epidermis. On their way up and out, the hair cells become keratinized. By the time you see it, the hair is mostly completely flattened, dead cells full of keratin. The curvature of the follicle causes the shape of the exposed hair, ranging from tight curls to stick straight.

## IT WILL MAKE YOUR HAIR STAND ON END

Each hair follicle has a tiny *arrector pili* muscle. When these smooth muscles contract, the previously bent hair shafts become erect. This movement causes the epidermis to be pushed to the side, and a goose bump appears. When you're cold or frightened, the sudden tightening of the arrector pili muscle causes the hair to rise up, and as a result, air is trapped between the hair and the skin. When you're cold, the trapped air acts as insulation — the same way a coat keeps you warm. When you're frightened, the hairs standing on end serve to make you look scary to whatever is scaring you, a form of signaling used by many mammals.

Hair goes through cycles of growth and dormancy. When the cells of the papilla begin to divide again, the hair falls out. This is why the length of your hair maxes out. The hairs on your head live about three to four years before you shed them, and eyelashes live about three to four months before falling out. People don't go bald overnight. Baldness (called *alopecia*) occurs when follicles go dormant do not reactivate.

## Nailing nails

Your fingernails and toenails lie on a *nail bed* (not to be confused with a bed of nails). At the back of the nail bed is the *nail root.* Just like skin and hair, nails start growing near the blood supply that lies under the nail bed, and the cells move outward at the rate of about 1 millimeter per week. As they move out over the nail bed, they become keratinized. (See Figure 4-2.)

At the bottom of your nails is a white, half-moon-shaped area called the *lunula.* (*Lun-* is the Latin root for moon, as in *lunar.*) The lunula is white because this is the area of cell growth. In the nail body, the nail appears pink because the blood vessels lie underneath the nail bed. But many more cells fill in the area of growth. This layer is thicker, and you see white instead of pink.

## Nothing's bland about glands

Glands in the skin make and secrete substances that are transported to your body's outer surface. The contraction of tiny muscles in the gland accomplishes this secretion. The two main types of skin glands are *sudoriferous glands* (sweat glands) and *sebaceous glands* (oil glands).

**FIGURE 4-2:**
A nail bed.

K·BORN

*Illustration by Kathryn Born, MA*

## Sudoriferous glands

Your body contains two types of sudoriferous glands. *Eccrine sweat glands* are distributed all over the skin. These glands open to the skin's surface, and when you're hot, they release sweat to reduce body temperature by a process of evaporative cooling. When the sweat, which is mostly water, evaporates, it takes the heat with it.

*Apocrine sweat glands* start to develop during puberty and connect to the hair follicles of the armpits and groin. Apocrine sweat contains a milky white substance and may also contain *pheromones,* chemicals that communicate information to other individuals by altering their hormonal balance. (Some research has indicated that the apocrine secretions of one woman can influence the menstrual cycle of other women who live with her.) Apocrine glands become active when you're anxious and stressed, as well as when you're sexually stimulated. Bacteria on the skin that digest the milky white substance produce unpleasantly odiferous byproducts.

The milk-secreting glands in mammary tissue are thought to have evolved from apocrine sweat glands.

## VITAMIN D

Sunlight has a nasty way of damaging skin, but you need a regular dose of sunlight to keep your bones healthy. Bones need vitamin D to develop healthy new bone cells. Insufficient vitamin D can lead to a condition called *rickets,* which results in soft, curved bones that can't support the body's weight.

Skin cells contain a molecule that's converted to vitamin D when struck by ultraviolet rays (UV) in sunlight. The vitamin D leaves the skin and goes through the bloodstream to the liver and kidneys, where vitamin D is converted to the hormone *calcitriol.* Calcitriol then circulates through the entire body and regulates the physiology of calcium and phosphorus, important minerals for the development and maintenance of healthy bones. (Scientists are discovering that vitamin D plays important roles in keeping your body healthy in other ways, too, such as mood regulation. To find out more, pick up *Vitamin D For Dummies* by Dr. Alan Rubin [Wiley].)

The use of sunscreen and sunblock products can prevent the UV radiation from reaching the skin cells and initiating vitamin D synthesis.

Just a few minutes of sunshine a day is all that's needed for your skin to make adequate amounts of vitamin D to keep your bones healthy. Anything beyond that amount of exposure is risky.

### Sebaceous glands

Sebaceous glands secrete an oily substance called *sebum* into hair roots. Besides wreaking havoc with teenage facial pores, sebum has physiological functions. It helps maintain your hair in a healthy state, which is important in regulating body temperature. It flows out along the hair shaft, coating the hair and the epidermis, forming a protective, waterproof layer. Sebum prevents water loss to the outside. Sebum also helps to protect you from infection by making the skin surface an inhospitable place for some bacteria.

In the watery environment of the amniotic sac, the human fetus produces a thick layer of sebum, called the *vernix caseosa.* Ear wax *(cerumen)* is a type of sebum produced by specialized cells in the ear canal.

# Your Skin Saving You

Your integumentary system participates in thousands of metabolic and homeostatic reactions, among them the thermoregulatory processes and the interaction of the skin and the nervous system.

## Controlling your internal temperature

Your skin plays an important role in homeostasis, specifically thermoregulation (see Chapter 2). It has mechanisms that increase body temperature when you're cold and decrease it when you're hot.

Your body is continuously converting energy from food into energy in the form of ATP (see Chapter 3). About 60 percent of food energy is converted to heat in the metabolic reactions that produce ATP, and more heat is given off in the reactions that use ATP, such as muscle contraction. This heat replaces the heat that your body loses continuously to the environment. (See Chapter 16 for an overview of ten basic chemistry and physics concepts.)

Human physiology employs specially adapted integumentary structures for thermoregulation. When your internal temperature rises above its set point, the blood vessels in the dermis dilate, dissipating heat from the blood to the environment through the epidermis. The sweat glands are activated, and heat escapes the body as the sweat evaporates.

When your skin is cold, the sweat glands aren't activated, thus slowing down evaporative cooling and raising your internal temperature by retaining the heat produced in metabolism. Blood vessels in the dermis constrict, limiting heat escape from the blood. If body temperature falls to about 97 degrees, shivering begins automatically, and these muscle contractions generate heat.

REMEMBER

Normal body temperature for a human is 97 to 100 degrees Fahrenheit (F). This is the optimal range for the very temperature-sensitive reactions of metabolism. A few degrees higher, and the body goes into convulsions. A few degrees lower, and metabolism gradually shuts down, sending the body into what may be its final sleep.

Other mammals, including other primates, rely on a more or less dense covering of hair on the epidermis to conserve body heat. During the evolution of the human species, natural selection favored the development of a very light coat of hair over most of the body. Turn to Chapter 17 for some information related to that evolution.

## Your skin is sensational

So how does your body know when it's cold or hot? How do you know when you get a cut or splinter? How can you tell the difference between being tickled with a feather and punched with a fist? Because the dermis contains nerve endings that serve as specialized receptors for hot and cold, touch, pressure, and pain. The receptors, such as *lamellated (Pacinian) corpuscles, tactile (Meissner's) corpuscles,* and

*Ruffini's corpuscles,* are sprinkled throughout the dermis and are connected to the nerves that run through the dermis and subcutaneous layer (hypodermis). See Chapter 7 for more about these receptors.

Not every square inch of skin contains all types of nerve endings. At one spot on your skin, you may sense light touch, while a few centimeters away, you may sense pressure; some spots sense cold, and some spots sense heat. These mixed messages connect with the network of nerves up to your brain, which does a pretty good job of making sense of the different kinds of information.

### Your skin is self-healing

Damage to the skin, such as a cut, triggers epidermal cells to divide more rapidly, trying to fill in the space. If the cut extends into the dermis or deeper, and a blood vessel is broken, the clotting blood will first cover the area allowing time for the epidermal cells to divide and fill in the space. A covering, natural (like a scab) or otherwise (like a Band-Aid) keeps nutrient-rich fluids from seeping out, as well as germs from getting in. Special cells called *phagocytes* will come in and clean up the debris. Inflammation is also an essential part of this process. The dilation of the blood vessels brings more resources to the area to speed the process.

If the cut is particularly deep, the skin cells will not be able to divide fast enough to fill it in. As a result, special cells in the dermis, called *fibroblasts,* will be triggered to produce collagenous fibers. These bulky fibers are what can lead to a scar.

# Pathophysiology of the Integument

Being in contact with the external environment, the integument inevitably encounters some nasty stuff: *pathogens* (disease-causing organisms like some bacteria, fungi, protists, and viruses), UV radiation, and harmful forces like fire, chemicals, and sharp objects. There are also hereditary pathophysiologies. Chances are that you've experienced — or known someone who has experienced — one of the conditions described in the following sections.

### Skin cancer

Many cases of skin cancer are related to UV exposure. Skin cancers are classified as a *melanoma* (a malignant or spreading type) or a *nonmelanoma* (a limited-to-one-portion-of-the-skin type).

>> **Basal cell carcinoma** is the most common form of skin cancer. UV radiation can cause a cancerous tumor to develop in the stratum germinativum. The immune system becomes increasingly unable to detect the tumor as it grows. This type of tumor is relatively easily removed and is usually easily cured.

>> **Squamous cell carcinoma** is a melanoma type that starts in the epidermis. Squamous cell carcinoma is more likely to spread to a nearby organ than basal cell carcinoma, and for every 100 people diagnosed with squamous cell carcinoma, one is likely to die from it.

>> **Malignant melanoma** starts in the melanocytes, the cells that produce melanin. The appearance of a malignant melanoma are almost black with irregular borders. These cancerous spots look like a spot on your garage floor where you spilled oil. Malignant melanoma occurs mostly in light-skinned people who have a history of severe sunburns, especially when they were kids. One in five people diagnosed with malignant melanoma die from it within five years.

# Dermatitis

*Dermatitis* is inflammation of the skin that takes the form of a rash that itches and burns. The inflammation can have many causes: infection, insect bites, irritation from chemicals, allergy, skin abrasion from shaving, or sunburn. A genetic pre-disposition is frequently involved, as in seborrheic dermatitis, or eczema, which often affects the scalp and hair as well as the skin of the hands, feet, face, or just about anywhere else.

Many people have an episode of dermatitis at some time in their lives, and some people suffer chronic dermatitis. Although the cause can vary, the treatment is often relatively simple: some combination of avoiding the cause or eliminating the infectious organism (fungus, for example) and applying a hydrocortisone cream to calm the inflammation and allow the skin to heal.

# Alopecia

*Alopecia* is the medical term for abnormal hair loss — that is, over and above the normal rate of hair loss caused by the follicles taking a break once in a while. Like dermatitis, the causes are quite various.

The most prevalent type is *androgeic alopecia*, or male pattern baldness. Androgeic alopecia is an inherited condition that affects about 25 percent of men before the age of 30 and two-thirds of all men before the age of 60. The condition is less common and less extreme in women. It can develop in older adults, resulting in an overall thinning of all the scalp hair rather than complete baldness.

*Alopecia areata* is a type of hair loss in which the immune system attacks hair follicles causing it to fall out in large patches. The root cause (sorry!) is unknown, but damage to the follicle is usually temporary. Alopecia areata is most common in people younger than 20, but children and adults of any age may be affected. Alopecia areata can be treated but not cured.

*Temporary alopecia* can result from a long list of causes, including stress, severe illness or surgery, nutritional deficiencies, treatment with certain drugs (notably chemotherapy agents for cancer, which attack all actively dividing cells), and certain medicines for arthritis, depression, heart problems, and high blood pressure. Use of some hair cosmetics or hair abuse (wearing the hair pulled back tightly over a long period of time) can harm the hair shaft or follicle. *Trichotillomania*, or compulsive hair pulling, can also eventually lead to alopecia.

## Nail problems as signs of possible medical conditions

Unhealthy nails can be a symptom of an underlying disease and can be of help in diagnosis. For example, bluish nail beds are caused by poor circulation. More specific indications include the following:

>> Brittle, concave (spoon-shaped) nails with ridges may indicate iron-deficiency anemia.

>> Nails separating from the nail bed can result from a thyroid disorder in which too much of a thyroid hormone is produced, such as Graves disease.

>> Black marks that look like tiny splinters under the nails can help diagnose respiratory disease or heart disease.

>> Hard, curved, yellow nails may indicate *bronchiectasis* (chronic dilation of bronchial tubes) and *lymphedema* (fluid retention in lymph glands).

Chapter **5**

# Scrutinizing the Skeletal System

I f you have any skeletons in your closet, now is the time to pull them out. We're not talking about your deep, dark secrets. Seriously! Actually looking at a model of a skeleton is the best way to figure out what's connected to what. If you don't have any skeletons in your closet, try your refrigerator. Observantly cutting up a chicken can show you a lot about bones and joints. And you'll get a head start on dinner!

The skeleton determines humans' general shape and size as a species, and also humans' very distinctive upright posture and bipedal gait. To get an overview of the skeleton, refer to the "Major Bones of the Skeleton" color plate in the center of this book.

In humans, as in all vertebrates, the skeleton is part of the *musculoskeletal system.* The other part, the muscular system, is the subject of Chapter 6.

The skeleton consists of all your bones, all the joints that connect your bones, and various kinds of fibrous tissue that cover, protect, and bind bones and joints together. In this chapter, we look at the special structures of these tissues and name some of the most important bones and joints. Other important functions of bone tissue, like mineral storage and blood cell production, are mentioned briefly or covered in detail in other chapters.

# Reporting for Duty: The Jobs of Your Skeleton

The structural functions of the skeletal system are these:

>> **Protection:** Bones and joints are strong and resilient. The rib cage provides a protected inner space for your more delicate internal organs. The vertebral column partially encases and protects the spinal cord, and the skull completely encases the brain.

>> **Movement:** The musculoskeletal system is a motion machine: The bones anchor the skeletal muscles and act as levers, the joints act as fulcrums, and muscle contraction provides the force for movement. (See Chapter 6 for information about muscles and muscle contraction.)

>> **Support:** The curved vertebral column supports most of your body's weight (see the section "Setting you straight on the curved spinal column" later in the chapter). The arches of your feet support the weight of your body in a different way (see the sidebar "Strong feet, firm foundation").

# Checking Out the Skeleton's Makeup

This section is about how your body builds the tissues of the bones and the joints and how they all fit together to protect, move, and support the entire body.

## Caring about connective tissue

The skeleton is made up mainly of three types of connective tissue: osseous (bone) tissue, cartilage, and fibrous connective tissue.

### Osseous tissue

*Osseous tissue* is physiologically very active, constantly generating and repairing itself, and has a generous blood supply all through it. Not only that, but bone makes a huge amount of "product for export," notably the very cells of the blood.

(Yes, new blood cells are made by bones — see Chapter 9.) Bone contains four specialized types of cells: osteocytes, osteoblasts, osteoclasts, and osteogenic cells. The skeletal system's functions depend on the functioning of these specialized cells in bone tissue.

**REMEMBER**

Keep in mind the difference between *bone tissue* and a specific, named bone. Both the *femur* (thigh bone) and the *humerus* (arm bone) contain bone tissue, but each bone has its own specialized configuration of the components of bone tissue.

We think of our bones as being hard like rocks, and they are, but the structure of the osseous tissue is quite complex. Figure 5-1 shows the parts of osseous tissue.

**FIGURE 5-1:**
Compact osseous
tissue structure.

*Illustration by Kathryn Born, MA*

The first thing you notice when looking at osseous tissue are the big holes. Then you see that everything seems to be arranged in circles around those holes. One set of these circles is called an *osteon;* these structures are repeated and glued together to form *compact bone.* The hole in the middle, the *Haversian* (or central) *canal,* creates space for the nerves and blood vessels to run throughout the bone.

Each ring encircling the central canal is called a *lamella.* These are formed as calcium compounds (such as calcium phosphate and calcium carbonate) are deposited into the matrix (the space between the cells). Unlike in other tissues where the matrix is fluid filled, bone cells, called *osteocytes,* are unable to move. They can be found in little caves called *lacunae.*

Also due to the rigid matrix, cells must get their nutrients from other cells. Osteocytes have armlike structures that reach through little tunnels through the matrix called *canaliculi.* Cells in the inner ring (lamella) have access to resources from the blood vessels in the central canal. They then pass these resources on to cells in the next ring through the canaliculi (tunnels). This continues, cell to cell, like a game of telephone.

## Cartilage

*Cartilage* is a firm but flexible tissue made up of mostly protein fibers. If you put your finger on the end of your nose and push gently, you can get a very good idea of the rubbery texture and flexibility of cartilage. Cartilage is the main component of joints.

Cartilage is less complicated in its structure than bone tissue, having fewer cells, fewer cell types, and little or no direct blood supply. However, among the functions of cartilage tissue is the building of new bone. The two types of cartilage in the skeletal system are hyaline cartilage and fibrocartilage.

>> *Hyaline cartilage* is the type that forms the septum of your nose. It also forms a portion of the very first version of the fetal skeleton. It's the most abundant type of cartilage in several kinds of joints — it's a major component of the freely movable ones called synovial joints.

>> *Fibrocartilage* is a fibrous, spongy tissue that acts as a shock absorber in the vertebral column (spine) and the pelvis.

Cartilage isn't generated and replaced as actively as bone, so cartilage gets by with fewer cells, and mature cartilage has no blood supply.

There is a third type of cartilage: *elastic cartilage.* Unlike the other two types, it contains numerous elastic fibers, making it much more flexible. You can find elastic cartilage in your epiglottis and your external ear.

## Fibrous connective tissue

*Fibrous connective tissue* (FCT) can be compared to the kind of packing tape that has fibers in it. FCT contains very few living cells and is composed mainly of protein fibers, complex sugars, and water.

FCT forms a structure called the *periosteum,* a protective sheet that covers bones. The collagen fibers in this covering intertwine with the collagen fibers of the *tendons* and *ligaments.* These cordlike structures connect a bone to another bone (ligaments), or a bone to a muscle (tendon).

The periosteum is said to be *continuous* with the ligaments and tendon, because there's no real separation between the "sheet" and the "cords." This prevents them from letting go of the bone.

## The structure of a bone

The structures called *bones* (the femur, the vertebrae, the finger bones) are made of bone tissue. (No surprise there, you say, but note that structures called *joints* are made of the tissue called *cartilage.*) It's important to remember that different individual bones have different forms of bone tissue.

Long bones such as your thighbone (femur) or forearm bone (radius) are the type of bones people usually think of first. And in fact, they make a good illustration of general anatomy and physiology of bone tissue (see Figure 5-2).

**FIGURE 5-2:**
Long bone
structure.

*Illustration by Kathryn Born, MA*

In cross section, bone is structured in concentric layers, that is, an outer layer surrounds a middle layer, which in turn surrounds an inner layer. In longitudinal section, a bone has two ends, mostly similar, and a long middle area, which has cells and tissues mostly different from the ends. The following list names and briefly describes cellular and material composition of the areas of a long bone:

>> **Compact (cortical) bone** (the outer layer) is a dense layer of cells in a hard matrix of protein fibers and compounds made of calcium and other minerals. This is the layer that gives bones their amazing strength. Compact bone tissue is described in detail in the "Caring about Connective Tissue" section, earlier in this chapter.

>> **Spongy (trabecular) bone** (the middle layer) is, like compact bone, a variety of cell types within a matrix of mineralized protein fibers. But spongy bone has a more open structure than compact bone, a physiological trade-off between strength and lightness. The matrix of spongy bone is not arranged in concentric circles; instead, it appears, well, spongy. Structures called *trabeculae,* which follow stress lines in the bone, act like braces, providing support while leaving large pockets of space. In adults, this space contains red marrow, described in the next subsection.

>> **The medullary cavity** is the inner layer of the shaft of a long bone (the *diaphysis*). The medullary cavity houses *bone marrow,* which comes in two varieties. There's *yellow marrow,* which is mostly fat (think butter), and *red marrow,* the site of *hematopoiesis,* the production of blood cells. In adults, most of the marrow in the medullary cavity is deactivated and, thus, yellow. The active, red marrow is found in spongy bone of the skull, ribs, vertebrae, pelvis, and sternum (breastbone). In infants, medullary cavities of the long bones are mostly filled with red marrow to meet the increasing demand for blood cells.

>> **Epiphysis** is the enlarged, knobby end of a long bone. It consists of an outer layer of compact bone overlying spongy bone. This is the site of bone elongation. Within the epiphysis, bone and cartilage tissue are intimately connected: As cartilage cells divide, cartilage morphs into bone tissue. This process continues from before birth until the bones reach their full adult size.

If blood cell levels drop too low — after blood loss, for instance — the yellow marrow can be reactivated and again perform hematopoiesis.

## Classifying bones

Bones come in different shapes and sizes. Appropriately, many bone type names match what they look like, such as flat bones, long bones, short bones, and irregular bones. Check out Table 5-1 for the differences among the four types of bones.

**TABLE 5-1** **Characteristics of Bone Types**

| Bone Type | Example Location in the Body | Characteristics |
|---|---|---|
| Flat | Skull, shoulder blades, ribs, sternum, pelvic bones | Like plates of armor, flat bones protect soft tissues of the brain and organs in the thorax and pelvis. |
| Long | Arms and legs | Like steel beams, these weight-bearing bones provide structural support. |
| Short | Wrists (carpal bones) and ankles (tarsal bones) | Short bones look like blocks and allow a wider range of movement than larger bones. |
| Irregular | Vertebral column, kneecaps | Irregular bones have a variety of shapes and usually have projections that muscles, tendons, and ligaments can attach to. |

# Bone Growth and Remodeling

When long bones develop in a fetus, they're formed from hyaline cartilage. The softer cartilage allows the fetus to bend into the poses that would make a yoga instructor beam with pride. The shape of the bone is determined by the shape of the cartilage, so it serves as a template. Calcium compounds are deposited onto the template, and the cartilage becomes calcified.

**TECHNICAL STUFF**

The terms *calcification* and *ossification* are often used synonymously to refer to the formation of bone. However, ossification is properly the formation of the whole tissue, whereas calcification is the actual formation and deposition of the calcium compounds.

Even at birth, the bones are not fully ossified. They continue to grow and develop well into the teenage years. This occurs via two types of ossification:

» **Intramembranous ossification:** Intramembranous ossification occurs in all bones; it's how short, flat, and irregular bones grow in size, and it's how long bones increase their width. A sheet of connective tissue is formed beneath the periosteum. Specialized bone cells called osteoblasts place the calcium compounds onto the fibers of the connective tissue, creating the trabeculae of spongy bone (calcification). When they're done building, osteoblasts become osteocytes. Later, the new spongy bone can be further developed into compact bone as new osteoblasts fill the empty space with calcium compounds until they've walled themselves in to a little cave — a lacuna.

» **Endochondral ossification:** To increase their length, long bones use *endo-chondral ossification.* In Figure 5-2, you see an *epiphyseal line* on each end. Rather than being osseous tissue, the epiphyseal line (or plate) is made of hyaline cartilage. When stimulated by growth hormone (see Chapter 8), the *chondrocytes* (cartilage cells) begin to copy themselves. The cells will then enlarge, which creates more space, thus lengthening the bone. Osteoblasts will then calcify the cartilage just as they did to the connective tissue during intramembranous growth.

When the long bones increase in length, you increase in height. Eventually, around age 18, the chondrocytes stop dividing. The entire epiphyseal line is *ossified* (turned into bone). This is commonly referred to as the growth plates being closed.

While you're likely done growing at this point in your life, your bones are definitely not done developing. This continues throughout life via a process called *remodeling.*

Because bones are constantly absorbing the forces placed on our bodies, the matrix gets damaged. The remodeling process allows the bones to retain their structural integrity by constantly replacing the weakened tissue with strong, healthy matrix. It works like road construction. The roads wear out with continued use and need to be refreshed. But you can't do all the roads at once or else people wouldn't be able to get places. So you do it in sections at a time, all the time. Luckily, bone remodeling doesn't cause the headaches that road construction does!

Before we can activate osteoblasts to build new tissue, we must first clear the area. Specialized bone cells called *osteoclasts* are responsible for this task. Osteoclasts secrete acid to break down the weakened matrix. This releases the calcium ions from their compounds, allowing them to be absorbed into blood flow. Then osteoblasts recycle the calcium, building new compounds to deposit and reform the matrix stronger than before.

The forgotten function of bones is to store calcium, an ion essential for both muscle contraction and nervous communication (see Chapters 6 and 7). When blood calcium levels drop too low, parathyroid hormone (PTH) is released and proceeds to the bones. The PTH then stimulates osteoclasts to resorb the calcium.

# The Axial Skeleton

The *axial skeleton* consists of the bones that lie along the midline (center) of your body, such as your vertebral column (backbone). An easy way to remember what bones make up the axial skeleton is to think of the vertebral column running down the middle of your body and then the bones that are directly attached to it — the thoracic cage (rib cage) and the skull.

The following sections give you a closer look at the main parts of the axial skeleton.

## Keeping your head up: The skull

Rather than one big piece of bone, like a cap that fits over the brain, the skull comprises the cranial and the facial bones.

A human skull (see Figure 5-3) is made up of the *cranium*, which is formed of several bones, and the *facial bones.* The facial bones contain cavities called the *sinuses*, which do have a purpose other than harboring upper respiratory infections.

**FIGURE 5-3:**
The human skull: cranium and facial bones.

*Illustration by Kathryn Born, MA*

## Cracking the cranium

The eight bones of your cranium protect your brain and have immovable joints between them called *sutures.* (These look a lot like the sutures, or stitches, that you may receive to close an incision or wound.) The bones of the cranium that are joined together by sutures include the following:

>> **Frontal bone:** Gives shape to the forehead and part of the eye sockets.

>> **Parietal bones:** Two bones that form the roof and sides of the cranium.

>> **Occipital bone:** Forms the back of the skull and the base of the cranium. The foramen magnum, an opening in the occipital bone, allows the spinal cord to pass into the skull and join the brain.

>> **Temporal bones:** Form the sides of the cranium near the temples. The temporal bone on each side of your head contains the following structures:

- **External auditory meatus:** The opening to your ear canal

- **Mandibular fossa:** Joins with the mandible (the lower jaw)

- **Mastoid process:** Provides a place for neck muscles to join your head

- **Styloid process:** Serves as an attachment site for muscles of the tongue and larynx (voice box)

>> **Ethmoid bone:** Contains several sections, called *plates*. Forms the medial (inside) part of the eye sockets and much of the nasal cavity.

>> **Sphenoid bone:** Shaped like a butterfly or a saddle (depending on how you look at it), the sphenoid forms the floor of the cranium and the back and lateral sides of the eye sockets (orbits). A central, sunken portion of the sphenoid bone called the *sella turcica* shelters the pituitary gland, which is very important in controlling major functions of the body. (See Chapter 8 for more on the pituitary gland.)

## Facing the facial bones

The bones that form facial structures are

>> **Lacrimal bones:** Two tiny bones on the inside walls of the orbits. A groove between the lacrimal bones in the eye sockets and the nose forms the *nasolacrimal canal.* Tears flow across the eyeball and through that canal into your nasal cavity, which explains why your nose "runs" when you cry.

>> **Mandible:** The lower jaw and the only movable bone of the skull.

>> **Maxillae:** Two bones that form the upper jaw, part of the hard palate (roof of your mouth), and bottom of the orbits.

>> **Nasal bones:** Two rectangular-shaped bones that form the bridge of your nose. The lower, movable portion of your nose is made of cartilage.

## THE BONE THAT FLOATS

The *hyoid* bone, a tiny, U-shaped bone that resides just above your larynx (voice box), anchors the tongue and muscles used during swallowing. However, the hyoid bone itself isn't attached to anything. It's the only bone in the body that does not articulate (connect) with another bone. The hyoid bone hangs by ligaments attached to the styloid processes of the temporal bones.

>> **Palatine bones:** Form the back portion of the hard palate and is the floor of the nasal cavity.

>> **Vomer bone:** Joins the ethmoid bone to form the nasal septum — that part of your nose that can be deviated by a strong left hook.

>> **Zygomatic bones:** Form the cheekbones and the lateral (outer) sides of the orbits.

## Sniffing out the sinuses

The sinuses allow air into the skull, making it much lighter. The air in your sinuses also gives resonance to your voice, which means that when you talk, the sound waves reverberate in your sinuses.

Several types of sinuses are named for their location:

>> **The frontal sinus** is a hollowed-out area in the frontal bone.

>> **Mastoid sinuses** drain into the middle ear.

>> **Maxillary sinuses** are large and within the bones of the upper jaw (the maxilla).

>> **Paranasal sinuses** consist of the frontal, sphenoidal, and ethmoidal sinuses, which, along with the maxillary sinuses, drain into the nose (*para* means "near"; *nas-* means "nose").

# Setting you straight on the curved spinal column

The spinal column (see Figure 5-4) begins within the skull and extends down to the pelvis. It's made up of 33 bones in all: 24 separate bones called *vertebrae* (singular, *vertebra*), plus the fused bones of the sacrum and the coccyx. Between each vertebra is an *intervertebral disc* made of fibrocartilage for shock absorption.

Your spinal column is the central support for the upper body, carrying most of the weight of your head, chest, and arms. Together with the muscles and ligaments of your back, your spinal column enables you to walk upright.

Cervical
curvature
C1 to C7

Thoracic
curvature
T1 to T12

Lumbar
curvature
L1 to L5

Sacral
curvature
S1 to S5

**FIGURE 5-4:**
The spinal
column, side
view.

*Illustration by Kathryn Born, MA*

**REMEMBER**

An important purpose of the vertebral column is to protect your spinal cord, the big data pipe between your body and your brain. Nearly all your nerves are connected, either directly or through networked branches, to the spinal cord, which runs directly into the brain through the opening in the skull called the *foramen magnum.* Turn to Chapter 7 for more information about the spinal cord.

If you look at the spine from the side, you notice that it curves four times: inward, outward, inward, and outward. The curvature of the spine helps it absorb shock and pressure much better than if the spine were straight. A curved spine also affords more balance by better distributing the weight of the skull over the pelvic bones, which is needed to walk upright. A curved spine keeps you from being top-heavy. Each curvature spans a region of the spine: cervical, thoracic, lumbar, and sacral. The number of vertebrae in each region and some important vertebral features are given in Table 5-2.

**TABLE 5-2**     **Regions of the Vertebral Column**

| Region | Number of Vertebrae | Features |
|---|---|---|
| Cervical | 7 | The skull attaches at the top of this region to the vertebrae called the *atlas*. |
| Thoracic | 12 | The ribs attach to this region. |
| Lumbar | 5 | Commonly referred to as the small of the back, it takes the most stress. |
| Sacral | 5 (fused into one; the sacrum) | The sacrum forms a joint with the hipbones and the last lumbar vertebra. |
| Coccygeal | 4 (fused into one; the coccyx, also called the tailbone) | The coccyx absorbs the shock of sitting. |

The vertebral column also provides places for other bones to attach. The skull is attached to the top of the cervical spine. The first cervical vertebra (abbreviated C-1; "C" for cervical, "1" for first) is the *atlas*, which supports the head and allows it to move forward and back (for example, the "yes" movement). The second cervical vertebra (C-2) is called the axis, and it allows the head to pivot and turn side to side (that is, the "no" movement).

**TIP**

You can differentiate these two important bones by recalling the Greek story about Atlas, who held the world on his shoulders. Your atlas holds your head on your shoulders. To remember the number of vertebrae in each region, think: breakfast, lunch, and dinner.

## Being caged can be a good thing

**REMEMBER**

The rib cage (also called the *thoracic cage*) consists of the thoracic vertebrae, the ribs, and the sternum (see Figure 5-5). The rib cage is essential for protecting your heart and lungs and for providing a place for the pectoral girdle (scapulae and clavicles) to attach.

You have 12 pairs of *bars* in your cage. Some of your ribs are *true* (7), some are *false* (3), and some are *floating* (2). All ribs are connected to the bones in your back (the thoracic vertebrae). In the front, true ribs are connected to the sternum (breastbone) by individual *costal cartilages* (*cost-* means "rib"); false ribs are connected to the *sternum* by fused *costal cartilage*. The last two pairs of ribs are called *floating ribs* because they remain unattached in the front. The floating ribs give protection to abdominal organs, such as your kidneys, without hampering the space in your abdomen for the intestines.

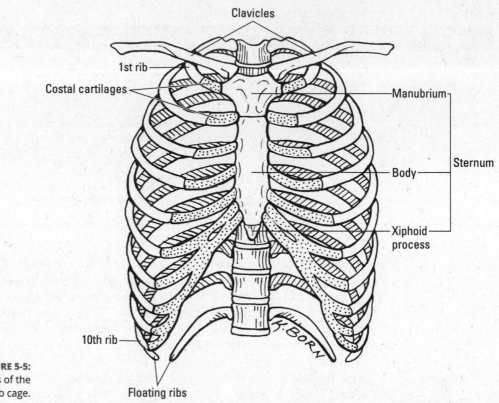

Clavicles

1st rib

Costal cartilages

Manubrium

Body

Sternum

Xiphoid
process

10th rib

Floating ribs

K.BORN

**FIGURE 5-5:**
The bones of the
rib cage.

*Illustration by Kathryn Born, MA*

The sternum (breastbone) has three parts: the *manubrium,* the *body,* and the *xiphoid* (pronounced *zi*-foid) *process.* The notch that you can feel at the top center of your chest, in line with your collarbones (the *clavicles*), is the top of the manubrium. The middle part of the sternum is the body, and the lower part of the sternum is the xiphoid process.

# The Appendicular Skeleton

The *appendicular skeleton* is made up of the bones and joints of the *appendages* (upper and lower limbs) and the two *girdles* that join the appendages to the axial skeleton. We describe each of these categories in the following sections.

# Wearing girdles: Everybody has two

**REMEMBER**

The word *girdle* is a verb than means "to encircle." It has nothing to do with that funny undergarment all polite women wore in the early 20th century.

The body contains two girdles: the *pectoral girdle*, which encircles the vertebral column at the top, and the *pelvic girdle*, which encircles the vertebral column at the bottom. The girdles serve to attach the appendicular skeleton to the axial skeleton.

The pectoral girdle consists of the two *clavicles* (collarbones) and the two triangle-shaped *scapulae* (shoulder blades). The scapulae provide a broad surface to which arm and chest muscles attach. Refer to the "Major Bones of the Skeleton" color plate to see the individual parts of the pectoral girdle.

The clavicles are attached to the sternum's *manubrium*. Significantly, this is the only point of attachment of the pectoral girdle and the axial skeleton. Because of this relatively weak attachment, the shoulders have a wide range of motion but are prone to dislocation.

The pelvic girdle (see the pelvis in Figure 5-6) is formed by the hipbones (called *coxal bones*), the *sacrum,* and the *coccyx* (tailbone). The hipbones bear the weight of the body, so they must be strong.

The hipbones (coxal bones) are formed by the *ilium*, the *ischium*, and the *pubis.* The ilia are what you probably think of as your hipbones; they're the large, flared parts that you can feel on your sides. The part that you can feel at the tip of the ilium is the *iliac crest.* In your lower back, the ilium connects with the vertebral column at the sacrum; the joint that's formed is appropriately called the *sacroiliac joint* — a point of woe for many people with lower back pain.

## WHY WEIGHT-BEARING EXERCISE IS GOOD FOR YOU

You probably know that aerobic exercise and lifting weights are good for your heart and your muscles, but did you realize it benefits your bones, too? Exercise — especially weight-bearing exercise (such as exercises that use the hips and legs, like walking, running, bicycling, and weight lifting) stimulates remodeling. More remodeling leads to stronger, denser bones. Thus, exercise staves off osteoporosis, which is the loss of bone density and the weakening of bones. Exercising the muscles also exercises the bones and joints, which maintains flexibility and strength. Nothing beats a firm foundation!

Pelvic girdle

Iliac crest

Ilium

Sacroiliac joint

Sacrum

Coccyx

Pubic symphysis

Pubis

Ischium

Pubic arch

**FIGURE 5-6:**
The bones
of the pelvis.

*Illustration by Kathryn Born, MA*

The *ischium* is the bottom part of your hip. You have an ischium on each side, within each buttock. You're most likely sitting on your *ischial tuberosity* right now. These parts of your hips are also called the *sitz bones* because they allow you to sit. The *ischial tuberosity* points outward and is the site where ligaments and tendons from the lower limbs attach. The *ischial spine* — which is around the area where the ilium and ischium join — is directed inward into the pelvic cavity. The distance between a woman's ischial spines is key to her success in delivering an infant vaginally (see Chapters 14 and 15); the opening between the ischial spines must be large enough for a newborn's head to pass through.

The *pubis* bones join the right and left hip bones together. They are joined together by a piece of fibrocartilage called the *pubic symphysis*. Pelvic floor muscles attach to the pelvic girdle at the pubis.

## WHY WOMEN HAVE BIGGER HIPS THAN MEN

Okay. You know it's true. Women aren't built like men. Most men tend to be straight up and down with few curves. Women, on the other hand, are hourglasslike — their hips tend to be wider than men's hips. In a woman, the iliac bones flare wider than they do in a man. The pubic arch is at an obtuse angle (greater than 90 degrees) where in males it is acute (less than 90 degrees). And the *lesser* (or true) *pelvis* of a woman — the ring formed by the pubic bones, ischium, lower part of the ilium, and the sacrum — is wider and more rounded. The lesser pelvis of a man is shaped somewhat like a funnel.

These differences in anatomy have a physiological purpose: Women have babies, and when those babies are ready to be born, they need to pass through a woman's true pelvis without getting stuck. Other differences also relate to giving birth: The sacrum in women is wider and tilted back more so than in men; the coccyx of women moves easier than it does in men. These two features allow a little more "give" when a baby is passing through the pelvis. When a woman is pregnant, she creates a hormone called *relaxin* that allows the ligaments connecting the pelvic bones to relax a little. Thus, the bones can spread farther apart and flex a bit during delivery.

All these female characteristics are wonderful for giving birth, but the bones don't always return to their original positions as readily. The hips tend to stay a bit wider after giving birth. Maybe the bones staying loose and spreading out is a physiological "reward," making delivery of another infant even easier. Too bad female bodies don't know when that last infant has been delivered so the hips can go back to their pre-pregnant sizes!

## Going out on a limb: Arms and legs

**REMEMBER**

Your arms and legs are *limbs* or appendages. The word *append* means to attach something to a larger body. Your appendages are attached to the axial skeleton by the girdles (see the preceding section).

### Giving you a hand with hands (and arms and elbows)

Your upper limb or arm is connected to your pectoral girdle. The bones of your upper limb include the *humerus* (arm), the *radius* and *ulna* (forearm), the *carpals* of the wrist, and the hand, which is made up of the *metacarpal bones* and the *phalanges* (refer to Figure 5-7).

Clavicle

Scapula

Humerus

Radius — — Ulna

1 —— Carpals

5 — Metacarpals

Proximal—
Middle —
Distal —

Phalanges

**FIGURE 5-7:**
The bones of
the upper limb
and hand.

*Illustration by Kathryn Born, MA*

The head (ball at the top) of the humerus connects to the scapula at the *glenoid cavity*. Muscles that move the arm and shoulder attach to the *greater* and *lesser tubercles*, two points near the head. Between the greater and lesser tubercles is the *intertubercular groove*, which holds the tendon of the biceps muscle to the humerus bone. The humerus also attaches to the deltoid muscle of the shoulder at a point about halfway down called the *deltoid tuberosity*. The muscle attached to the deltoid tuberosity allows you to raise and lower your arm.

The bones of the forearm attach at the elbow end of your humerus in four different spots:

>> **Capitulum:** Two knobs that allow the radius to articulate (join) with the humerus.

>> **Trochlea:** A pulley-like feature on the humerus that lies next to the capitulum and allows the *trochlear notch* of the ulna to articulate with the humerus.

>> **Coronoid fossa:** Depression in the humerus that accepts a projection of the ulna bone (called the *coronoid process*) when the elbow is bent.

>> **Olecranon fossa:** Depression in the humerus that accepts a projection of the ulna (called the *olecranon process*) when the arm is extended. Fitting, isn't it?

REMEMBER

The radius is the bone on the thumb side of your forearm. When you turn your forearm so that your palm is facing backward, the radius crosses over the ulna so that the radius can stay on the thumb side of your arm. The radius is shorter but thicker than the ulna. The head of the radius is flat like the head of a nail. The ulna is long and thin, and its head is at the opposite end of the bone compared with the head of the radius.

Both the radius and ulna connect with the bones of the wrist. The wrist contains eight short bones called the *carpal bones.* The ligaments binding the carpal bones are very tight, but the numerous bones allow the wrist to flex easily. The eight carpal bones are arranged in two rows. The proximal row (furthest from your fingertips) contain the scaphoid, lunate, triquetrum, and pisiform (from thumb to pinky — note that *pisiform* and *pinky* both start with *p*). The distal row (also from thumb to pinky) contain the trapezium, trapezoid, capitate, and hamate.

The palm of your hand contains five bones called the *metacarpals*. When you make a fist, you can see the ends of the metacarpals as your knuckles. Your fingers are made up of bones called the *phalanges;* each finger has three phalanges (phalanx is singular): the *proximal phalanx,* which joins your knuckle, the *middle phalanx,* and the *distal phalanx,* which is the bone in your fingertip. The thumb, though, only has two phalanges, so some people like to argue that it's not considered a true finger. So you may have eight fingers and two thumbs or ten fingers depending on how you look at it. Regardless, on each hand the thumb is referred to as the first digit.

## Getting a leg up on your lower limbs

Your lower limb consists of the *femur* (thigh bone), the *tibia* and *fibula* of the leg, the bones of the ankle *(tarsals)*, and the bones of the foot (*metatarsals* and *phalanges*; refer to Figure 5-8).

FIGURE 5-8:
The bones of the
lower limb (a),
ankle, and
foot (b).

*Illustration by Kathryn Born, MA*

The term *phalanges* refers to the finger bones *and* the toe bones.

**REMEMBER**

The femur is the strongest bone in the body; it's also the longest. The head of the femur fits into a hollowed out area of the hip bone called the *acetabulum.* In women, the acetabula (plural) are smaller but spread farther apart than in men. This anatomic feature allows women to have a greater range of movement of the thighs than men. The *greater* and *lesser trochanters* of the femur are surfaces to which the muscles of the legs and buttocks attach. Trochanters are large processes found only on the femur. The *linea aspera* is a ridge along the back of the femur to which several muscles attach.

The femur forms the knee along with bones of the lower leg. The *patella* (commonly known as the kneecap) articulates with the bottom of the femur. The femur also has knobs (*lateral* and *medial condyles*) that articulate with the top of the tibia. The ligaments of the patella attach to the *tibial tuberosity*. The bottom, inner end of the tibia has a bulge called the *medial malleolus*, which forms part of the inner ankle.

The tibia, also called the shinbone, is much thicker than the fibula and lies on the medial (inside) portion of the leg. Although the fibula is thinner, it's about the same length as the tibia. The bottom, outside end of the fibula is the *lateral malleolus*, which is the bulge on the outside of your ankle.

Your foot is designed in much the same way as your hand. The ankle, or *tarsus*, which is akin to the wrist, consists of seven tarsal bones. However, only one of those seven bones is part of a joint with a great range of motion — the *talus*. The talus bone joins to the tibia and fibula and allows for the movements of your ankle. Beneath the talus is the largest tarsal bone, the *calcaneus*, which is the heel bone. The calcaneus and talus help to support your body weight. The remaining tarsals are the *cuboid* on the outside, the *navicular*, and the lateral, intermediate, and medial *cuneiforms*.

The instep of your foot is akin to the palm of your hand, and just as the hand has metacarpals and phalanges, the foot has metatarsals and phalanges. The ends of the metatarsals on the bottom of your foot form the ball of your foot. As such, the metatarsals also help to support your body weight. Together, the tarsals and metatarsals held together by ligaments and tendons form the arches of your feet. Your toes are also called phalanges, just like your fingers. And, just as your thumbs have only two phalanges, your big toes have only two phalanges. But, the rest of your toes have three: proximal, middle, and distal.

## STRONG FEET, FIRM FOUNDATION

The arches of your feet help to absorb the shock created by the pounding of your feet when walking or running, and they help to distribute your weight evenly to the bones that bear the brunt of it: the *calcaneus* (heel) and *talus* (ankle) in each foot. The heel, ankle, and metatarsals bear a significant amount of your body weight and are thus called *weight-bearing bones*. If the ligaments and tendons that form those arches along with the tarsals and metatarsals weaken, the arches can collapse, resulting in flat feet. A person with flat feet is more prone to damage to the feet bones (such as the metatarsals) because of increased pressure placed on those bones. Examples of damage include bunions, which is a painful displacement of the first metatarsal (the big toe), and heel spurs, which are bony outgrowths on the calcaneus that cause pain when walking. Flat feet also can contribute to knee pain, hip pain, and lower back pain.

# Joints and the Movements They Allow

A *joint*, or *articulation*, is a connection between two bones. Some joints move freely, some move a little, and some never move. This section tells you about the different joint structures and the movements they allow.

## Categorizing the types of joints

Joints, which vary greatly in their size and shape, can be classified by the amount of movement they permit or by their structure.

### Structural groupings

Joints fall into three categories based on the type of connective tissue present where the bones meet:

>> **Fibrous:** Bones held tightly together by dense connective tissue containing numerous collagen fibers

>> **Cartilaginous:** Bones held together with either hyaline or fibrocartilage

>> **Synovial:** Bones, which are lined with hyaline cartilage, are held together by a connective tissue capsule

### Immovable joints

*Synarthroses* are joints that don't move, such as those between the bones of the skull. A thin layer of fibrous connective tissue, called a *suture*, joins them together. The sutures in the cranium are named as follows:

>> **Coronal suture:** Joins the parietal bones and the frontal bone

>> **Lambdoidal suture:** Joins the parietal bones and the occipital bone

>> **Sagittal suture:** Between the parietal bones

>> **Squamosal sutures:** Between the parietal and temporal bones

Most fibrous joints are synarthroses.

### Slightly movable joints

*Amphiarthroses* are slightly movable joints connected by fibrocartilage or hyaline cartilage. Examples include the *intervertebral disks,* which join each vertebrae and allow slight movement of the vertebrae.

Most cartilaginous joints are amphiarthroses.

## Freely movable joints

*Diarthroses* are joints that are freely movable. The numerous types of diarthroses are shown in Table 5-3.

All diarthroses are also *synovial joints.* The *joint capsule* creates a cavity between the two connecting bones which is filled with *synovial fluid,* to help lubricate and cushion the joint. The ends of the bones are cushioned by hyaline cartilage and the range of movement allowed depends greatly on their shape.

**TABLE 5-3** **Types of Diarthroses (Synovial Joints)**

| Type of Joint | Description | Movement | Example |
|---|---|---|---|
| Ball-and-socket joint | A joint in which the ball-shaped head of one bone fits into a depression (socket) in another bone | Circular movements; can move in all planes, and rotation is possible. | Shoulder, hip |
| Condyloid joint | A joint in which the oval-shaped condyle of one bone fits into the oval-shaped cavity of another bone | Can move in all planes, but can't rotate. | Knuckles (joints between metacarpals and phalanges) |
| Gliding joint | A flat or slightly curved surfaces joint | Sliding or twisting; movement in two planes. | Joints between carpal bones (wrist) and between tarsal bones (ankle) |
| Hinge joint | A joint in which a convex surface joins with a concave surface | Up and down motion in one plane; can bend (flex) or straighten (extend). | Elbow, knee |
| Pivot joint | A joint in which a cylinder-shaped projection on one bone is surrounded by a ring of another bone and ligament | Rotation is only movement possible. | Joint between radius and ulna at elbow and joint atlas and axis at top of vertebral column |
| Saddle joint | A joint in which each bone is saddle shaped and fits into the saddle-shaped region of the opposite bone | Many movements are possible; can move in different planes but can't rotate. | Joint between carpal and metacarpal bones of the thumb |

# Knowing what your joints can do

You know that certain types of joints can perform certain kinds of movements. The movement of a body part — say, raising your hand — often has an opposing movement to return it to its original position, like putting your hand down in frustration when you don't get called on. Here's a quick overview of those special movements:

>> **Abduction:** Moves a body part to the side, away from the body's middle. When you make a snow angel and you move your arms and legs out and up, that's abduction.

>> **Adduction:** Moves a body part from the side toward the body's middle. When you're in snow angel position and you move your arms and legs back down, that's adduction — you're "adding" your body back together.

>> **Flexion:** Decreases the joint angle. When you flex to show off your biceps, you move your forearm to your arm, decreasing the angle at the elbow.

>> **Extension:** Makes the angle larger. Returning your arm from the flexed position increases the angle and the elbow and is, thus, extension. Hyperextension occurs when the body part moves beyond a straight line (180 degrees) like tilting your head back in exasperation.

>> **Elevation:** The upward movement of a body part, such as shrugging your shoulders.

>> **Depression:** The downward movement of a body part, such as the downward movement after shrugging your shoulders.

>> **Eversion:** Happens only in the feet when the foot is turned so the sole is facing outward.

>> **Inversion:** Happens only in the feet when the foot is turned so that the sole is facing inward.

>> **Supination:** Happens only in the arm, when the forearm is rotated to make the palm face upward or forward (think about holding a bowl of soup).

>> **Pronation:** Happens only in the arm, when the forearm is rotated to make the palm face downward or backward.

>> **Rotation:** The movement of a body part around its own axis, such as shaking your head to answer, "No." The partnered motions are *medial rotation* (movement toward the midline) and *lateral rotation* (movement away from the midline).

>> **Circumduction:** The movement of a body part in circles, like doing arm circles in gym class.

# Pathophysiology of the Skeletal System

Bones and joints are very strong, but they're prone to injuries, the effects of aging, and disease, just like any other body part. This section gives you some information on a few of the most common problems that occur in bones and joints.

## Abnormal curvature

Abnormal curvatures of the spine can cause plenty of pain and can lead to several problems. When the curve of the lumbar spine is exaggerated, the abnormal condition is *lordosis,* more commonly known as *swayback.* The lumbar spine of a pregnant woman becomes exaggerated because the woman needs to balance the pregnant belly on her frame. However, sometimes the curve remains after pregnancy, when weakened abdominal muscles fail to support the lumbar spine in its normal position. Developing the habit of holding the abdominal muscles in (rather than letting it all hang out, so to speak) helps to strengthen the body's center and prevent swayback. Losing the beer belly helps, too.

Older men and women sometimes develop a condition called *kyphosis* (commonly known as *hunchback*), an abnormally curved spine in the thoracic region. Normal degeneration and compression of the vertebrae tends to straighten the cervical and lumbar regions of the spine and push out the thoracic vertebrae, thus causing kyphosis. Osteoporosis (see the next section) amplifies this.

You may recall being checked for *scoliosis* during junior-high gym class. The reason for that inspection is because scoliosis (abnormal lateral curvature) first becomes obvious during the late childhood/early teen years — just when people are most self-conscious. Normally, when you look at the spine from the back, it appears to be straight — the curvature is evident only when you view the spine from the side. However, in people with scoliosis, the spine curves side to side (laterally) and looks S-shaped when viewed from the back.

## Osteoporosis

*Osteoporosis* is a disease in which bones become fragile, progressively and painlessly. To some extent, the process is inevitable with age, but when too much bone density is lost and small fractures appear, you have osteoporosis. The continuous process of bone resorption (osteoclasts breaking down the matrix) continues, but the osteoblasts (bone-building cells) become less and less active, so more bone is lost than replaced. Osteoporosis occurs most often in postmenopausal women because they lose the protective effect of estrogen on the bones.

Osteoporosis affects all bones, but of special concern are fractures of the hip and spine. A hip fracture almost always requires hospitalization and major surgery. It can impair a person's ability to walk unassisted and may cause prolonged or permanent disability or even death. Spinal or vertebral fractures also have serious consequences, including loss of height, severe back pain, and deformity.

# Cleft palate

A *cleft palate* is a relatively common birth defect that occurs when the *palatine bones* (a pair of the facial bones) or the *maxilla bones* fail to fuse during fetal development. This defect creates a problem in which the nasal cavity and oral cavity are open to each other. This problem can affect the palatine bones only or can be part of a syndrome of development problems. Cleft palate, also called *hare-lip*, is treated with surgery, usually when the child is very young.

# Arthritis

*Arthritis* is a name for any of numerous conditions characterized by inflammation of the joints. The inflammation is painful in itself, and it also makes movement difficult and painful. The chronic inflammation can eventually erode the joint's tissues (bone and cartilage). Treatment consists of controlling pain, reducing inflammation, and slowing the progress of joint damage.

Arthritis conditions are closely associated with immunity: Inflammation is a normal response of the immune system, but chronic inflammation of the joints is pathophysiological. Several arthritis conditions are autoimmune disorders. (See Chapter 13 for a discussion of autoimmunity.) Here are the common forms of arthritis:

>> **Osteoarthritis (OA)** is the most common form. As the joints age and the ravages of normal use accumulate, low-level inflammation sets in. Eventually, the inflammation causes the joint's cartilage to become thinner and lose elasticity. It can affect people at any age. Many people develop some osteoarthritis of the finger joints in late middle age.

>> **Rheumatoid arthritis (RA)**, an autoimmune condition, starts with inflammation of the *synovium* (joint lining). Later, often years later, the inflamed cells of the synovium begin to produce enzymes that actively destroy both bone and cartilage, restricting movement and increasing pain further.

>> **Juvenile arthritis (JA)**, the most common form affecting children under 16, is an autoimmune condition. Most likely, JA is several different autoimmune conditions that vary in the number of affected joints and the age of onset.

As with other forms of arthritis, the symptoms are inflammation, joint pain, and stiffness. Sometimes, JA causes the limbs to grow to different lengths.

>> **Ankylosing spondylitis** (**AS**) affects the spine and the sacroiliac joints. The severity of the pain and inflammation varies, but in its worst form, the chronic inflammation can cause the spine to fuse into a rigid, brittle column, prone to fracture. The eyes, heart, lung, and kidneys can also be affected.

>> **Gout** is a form of arthritis caused by crystallized deposits of uric acid in the joints. Think "sand in the gears." As the uric acid crystals fill the joints, they damage cartilage, synovial membranes, tendons, and even the muscles adjoining the bone. Complications include kidney stones, nerve damage, and circulatory problems. Drugs are available that reduce the amount of uric acid in the blood, preventing its deposition in the joints.

# Fractures

When bones absorb more force than they can handle, they break. Fractures are classified by the shape of the break, whether or not it spans the whole bone, and if breaks through the skin (called *compound* if it does; *closed* if it does not). Not all fractures are visible lines, though. Compression fractures, where the bone crumples but doesn't literally break, are a common type of pathological fracture — those that have their root in disease (rather than trauma).

The body has its own fracture-repair process similar to remodeling, but key to proper healing is *reduction* (the correct realignment of the bone) and *immobilization* (which is maintained by applying casts or other equipment, such as plates or screws).

Chapter **6**

# Muscles: Setting You in Motion

**M**uscle tissues always have work to do. They pull things up and push things down and around. They move things inside you and outside you. And they move you, too, of course. They work together with all the other systems in your body, but more than any other organ system, the muscular system specializes in movement: of food and air into and out of your body; of the circulation of your blood within your body; of different parts of your body relative to one another, as when you change position; and of your body through space — what you normally think of as "movement."

All muscle tissue is strong. Most is enduring, some of it astoundingly so. Its cells are crowded with *mitochondria*, thousands of little factories constantly turning out molecules of *ATP*, a refined fuel. The muscle cells use the fuel to manufacture strong and flexible proteins, with which they build and repair themselves and do work.

All the work that muscle tissues do is done by the coordinated contraction and release of millions of *sarcomeres*, tiny structures within the muscle cells. Muscle activity accounts for most of the body's energy consumption.

In this chapter, we give you an overview of some of the things muscles do and how they do it, and then we name the muscles. At the end of the chapter, we list some common muscle ailments, one of which you've probably experienced.

# Functions of the Muscular System

This section focuses mainly on the functions of the *skeletal muscles*, the muscles that move your bones. The skeletal muscles comprise a substantial portion of your body mass, and most of what you eat goes to fuel their metabolism. In this section, you'll find out something about what they do with all that energy.

## Supporting your structure

Muscles are attached to bones on the inside of your body and skin on the outside, with various types of connective tissue between the layers. Thus, they hold your body together. Along with your skin and your skeleton, your muscles shield your internal organs from injury from impact or penetration.

Moreover, and don't take this personally, but your body is heavy. As it does with everything else, gravity pulls your weight downward (toward the planet's center). But gravity doesn't only pull on the soles of your feet — it pulls on all your weight. If gravity had its way, you'd be lying on the floor right now. Your muscles pull your weight up ("oppose" the pull of gravity) and hold you upright. Gravity is a relentless cosmic force, and eventually, gravity will win. But while you're still fighting, your muscles need fuel and rest.

## Moving you

Contracting and releasing a muscle moves the bone it's attached to relative to the rest of the body. The movement of the bone, in turn, moves all the tissue attached to it through space, as when you raise your arm. Certain combinations of these types of movements move the entire body through space, as when you walk, run, swim, skate, or dance.

Muscle contraction is responsible for little movements, too, like blinking your eyes, dilating your pupils, and smiling.

# Poised positioning

A very close interaction outside of your conscious control between some muscle cells and the nervous system keeps you not just upright, but in balance. Nerve impulses throughout the muscular system cause muscles to contract or relax to oppose gravity in a more subtle way when, say, you're shifting your weight from one side to the other as you step. This interaction is called *muscle tone,* and it's what is enabling you to hold your head up right now.

When you step down a steep incline in rough terrain, your muscle tone brings your abs and your back muscles into action in a different way than when you step across your living room rug. The mechanisms of muscle tone may move your arms up and away from your body to counterbalance the pull of gravity with an accuracy and precision you could never calculate cognitively. Below your conscious level, the mechanisms of muscle tone are active every minute of every day, even when you're asleep.

Muscle tone relies on *muscle spindles* — specialized muscle cells that are wrapped with nerve fibers. (See the "Skeletal muscle" section later in this chapter for information about spindle fibers.) The central nervous system stays in contact with the muscles through the muscle spindles. (Turn to Chapter 7 for the structures of the central nervous system.) Spindles send messages about your body position through the spinal cord to the brain; to initiate the fine adjustments, the brain sends signals through the spinal cord and nerves to the muscle spindles about which muscles to contract and which to release.

# Maintaining body temperature

Muscle contributes to *homeostasis* (see Chapter 2) by generating heat to balance the loss of heat from the body surface. Muscle contraction uses energy from the breakdown of ATP and generates heat as a byproduct. Shivering is a series of muscle contractions that generate extra heat to keep your temperature up in cold situations. If the heat generated by the muscles raises body temperature too high, other thermoregulatory processes, such as sweating, are activated.

# Pushing things around inside

The other two types of muscle tissues, smooth muscle and cardiac muscle, have their own important functions, discussed in other chapters. Here, we give you an overview of some of the muscles that keep things moving within your body, all without any thought from you.

## Cardiac muscle

The cardiac muscle that makes up the walls of the heart contracts rhythmically, pumping the blood into the arteries. It hits with so much force that the artery walls briefly stretch — detectable as your pulse. However, there must be enough force from the blood pushing on the walls by the time it reaches the capillaries to push out the plasma. This is why blood pressure is so important. Smooth muscle in the walls of the arteries help control this by contracting to constrict the vessel or relaxing to dilate it. Damage to this layer is one cause of *arteriosclerosis* (hardening of the arteries), which takes away the arteries' subtle control of blood pressure. See Chapter 9 for more about the heart and blood vessels.

## Diaphragm

The *diaphragm* is a skeletal muscle whose contraction and release forces air in and out of the lungs. Chapter 10 covers the respiratory system in all its breathtaking glory.

## Digestive smooth muscle

The organs of the digestive tract have walls containing smooth muscle that contracts in pulsating waves, pushing ingested material along. Think of this muscular lining as a conveyor belt on a disassembly line. Refer to Chapter 11 for details.

## Controlling release

*Sphincter muscles* are essentially valves: rings of smooth muscle that are fully contracted in their resting state, holding some material in one place, and then relaxing only briefly to allow the material to move through. You find sphincters at various places in the digestive system (refer to Chapter 11), from the very beginning to the very end, and in other parts of the body as well.

Most sphincters aren't under conscious control. Two of them — the urinary sphincter, which holds urine in the bladder, and the anal sphincter, which holds feces in the colon — come under conscious control usually at around 2 years of age. This control allows the release of these bodily wastes under culturally appropriate circumstances. Its acquisition is considered a milestone in infant development.

By the way, human males have much stronger urinary sphincter muscles than do females, meaning that they can retain about twice as much urine in the bladder (up to 800mL or 1.69 pints) for twice as long. Kindly keep that in mind, fellas, when you're on a road trip with girls.

# Talking about Tissue Types

A "muscle tissue type" is not the same as "a muscle." Your left bicep is a muscle; in all, you have hundreds of named muscles. (Refer to the "Naming the Skeletal Muscles" section later in the chapter and to the "Muscular System" color plate in the center of the book.) There are only three muscle tissue types: *skeletal muscle tissue, cardiac muscle tissue,* and *smooth muscle tissue.*

## Defining unique features of muscle cells

Your muscle tissue is made up of cells that are different from the other cells of your body. These cells are so unique that they're even different from each other, based on the type of muscle tissue they belong to. The three muscle types are distinguishable anatomically by their characteristic cells and structures and physiologically as *voluntary* or *involuntary.*

Muscle cells feature these characteristics:

>> **Single or multiple nuclei:** Cardiac muscle cells and smooth muscle cells have one nucleus apiece, like most other cells. Skeletal muscle cells (referred to as fibers) are *multinucleate,* meaning numerous nuclei are found within one cell membrane. Skeletal muscle cells don't grow extra nuclei; during the development of skeletal muscle tissue, numerous skeletal muscle cells merge into one large cell, and most of the nuclei are retained within one continuous cell membrane, along with most of the mitochondria.

>> **Striation:** Skeletal muscle is *striated,* meaning that, under a microscope, alternating light and dark bands are visible in the fiber (muscle cell). Striation is the result of the structures inside the skeletal muscle cells (see the "Skeletal Muscle" section later in the chapter) that carry out the mechanism of contraction called the *sliding filament model.* (See the "Getting a Grip on the Sliding Filament" section later in the chapter). Cardiac muscle cells are striated as well, and they also contract by a variation of the sliding filament model. Smooth muscle cells are not striated in appearance but do follow a version of the sliding filament model.

See Figure 6-1 to get an idea of how muscle cells and tissues are similar and different.

Muscle cells can also be categorized by the type of contraction they perform. Smooth and cardiac muscle cells are *involuntary,* meaning their contraction is initiated and controlled by parts of the nervous system that are far from the conscious level of the brain (the autonomic nervous system). You have no practical way to consciously control, or even become aware of, the smooth muscle contractions in your stomach that are grinding up this morning's muffin. The involuntary contractions that cause your *heartbeat* aren't even under the nervous system's control, as we discuss in Chapter 9.

Skeletal muscle tissue

Cardiac muscle tissue

Smooth muscle tissue

Striation

Muscle fiber (cell)

Nucleus

Cardiac muscle fiber (cell)

Striation

Intercalated disc

Nucleus

Smooth muscle fiber (cell)

Nucleus

**FIGURE 6-1:** Muscle cell and tissue types.

Illustration by Kathryn Born, MA

Skeletal muscle is classified as *voluntary* because you make a decision at the conscious level to move the muscle. At least you do sometimes — you decide to reach for a doorknob and turn it, for example, and your muscles carry out the command from your brain to do so.

But note this: If the doorknob is charged with static electricity, your arm pulls your hand away before you're even consciously aware of being zapped. This *somatic reflex arc* is still classified as voluntary movement, however, because it involves skeletal muscle, controlled by the somatic (voluntary) nervous system.

REMEMBER

Not everything in anatomy and physiology makes sense at first. Just remember that skeletal muscle is classified as voluntary.

Table 6-1 sums up the characteristics and classifications of muscle cells.

**TABLE 6-1** ## Cell Characteristics of Muscle Cells

| | Skeletal | Cardiac | Smooth |
|---|---|---|---|
| Multinucleate | Yes | No | No |
| Striated | Yes | Yes | No |
| Voluntary | Yes | No | No |

# Skeletal muscle

Skeletal muscle tissue is, essentially, bundles of fibers bundled together. Like fibrous material of every kind, skeletal muscle tissue gets its strength from assembling individual fibers together into strands, and then bundling and rebundling the strands. Two properties make this particular fibrous material very special: The strands are made of protein, and they renew and repair themselves constantly.

## At the cellular level

Individual muscle cells, which physiologists call *fibers*, are slender cylinders that sometimes run the entire length of a muscle. Each fiber (cell) has many nuclei located along its length and close to the cell membrane, which is called the *sarcolemma* in skeletal muscle fibers. Outside the sarcolemma is a lining called the *endomysium*, a type of connective tissue, which houses capillaries and nerves.

Muscle *spindles* are specialized skeletal muscle fibers that are wrapped with nerve fibers. Figure 6-2 shows how skeletal muscle is connected to the nervous system. Spindles are distributed throughout the muscle tissue and provide sensory information to the central nervous system. Motor neurons transmit impulses to trigger a muscle fiber to contract. Each fiber must be stimulated individually by a neuron at its *motor end plate*. However, a single motor neuron can stimulate numerous fibers, forming a *motor unit*. Large motor units (one neuron, numerous fibers) allow for gross motor skills like walking and lifting. Small motor units (one neuron, few fibers) provide fine motor skills like grasping and handwriting.

Cross-section of spinal cord
Motor neurons (efferent)
Muscle fibers
Perimysium
Sarcolemma of muscle cell
Epimysium
One myofibril
Fascicle
Skeletal muscle
Z line
Muscle cell nucleus
Sarcomere
Sarcoplasmic reticulum
T-tubule
Z line

*Illustration by Kathryn Born, MA*

**FIGURE 6-2:**
Anatomy of skeletal muscle tissue.

Bundled within the muscle fibers are *myofibrils* (refer to Figure 6-2). The myofibrils are composed of *sarcomeres,* which are distinct units arranged linearly (end to end) along the length of the myofibril. A sarcomere is the functional unit of muscle contraction. (Refer to the "Getting a Grip on the Sliding Filament Model" section later in the chapter for more on muscle contraction within sarcomeres.)

## At the tissue level

Muscle fibers are bound together into bundles called *fascicles.* Each fascicle is bound by a connective-tissue lining called a *perimysium.* Spindle fibers are distributed throughout each fascicle. The fascicles are then bound together to form a muscle, a discrete assembly of skeletal muscle tissue, like the *biceps brachii* (your biceps), with a connective-tissue wrapper called an *epimysium* holding the whole package together.

*Tendons* — ropy extensions of the connective tissue covering the skeletal bones — weave into the epimysium, holding the muscle firmly to the bone (see Chapter 5 for more on connectivity in the skeletal system). Muscles connect to other muscles using *aponeuroses,* a connective tissue similar to a tendon but broad and flat.

REMEMBER

How many ways can you say "fiber"? Anatomists need them all when they're talking about the muscular system. Make sure you're thinking at the right level of organization *(subcellular, cellular,* or *tissue)* when you see these terms: *filament, myofibril, fiber,* and *fascicle.*

## Working together: Synergists and antagonists

Groups of skeletal muscles that contract simultaneously to move a body part are said to be *synergistic.* The muscle that does most of the moving is the *prime mover.* The muscles that help the prime mover achieve a certain body movement are *synergists.* When you move your elbow joint, the bicep is the prime mover and the brachioradialis stabilizes the joint, thus aiding the motion.

*Antagonistic* muscles also act together to move a body part, but one group contracts while the other releases, a kind of push-pull. One example is flexing your arm. When you bend your forearm up toward your shoulder, your biceps muscle contracts, performing a *concentric contraction.* In the meantime, the triceps muscle in the back of your arm relaxes, performing an *eccentric contraction.* The actions of the biceps and triceps muscles are opposite, but you need both actions to allow you to flex your arm, which is why they are both, confusingly, referred to as contractions. Antagonistic actions lower your arm, too: The biceps relaxes, and the triceps contracts.

# Cardiac muscle

The heart has its own very special type of muscle tissue, called *cardiac muscle.* The cells (fibers) in cardiac muscle contain one nucleus (they're *uninucleated*) and are cylindrical; they may be branched in shape. Unlike skeletal muscle, where the fibers lie alongside one another, cardiac muscle fibers interlock, which promotes the rapid transmission of the contraction impulse throughout the heart. Cardiac muscle cells are striated, like skeletal muscle cells, and cardiac muscle contraction is involuntary, like smooth muscle contraction. Cardiac muscle fibers contract in a way very similar to skeletal muscle fibers, by a sliding filament mechanism (more on that in a minute).

Cardiac muscle tissue is on the job, day and night, from before birth to the moment of death. The cardiac muscle cells contract regularly and simultaneously hundreds of millions of times throughout your lifetime. When cardiac muscle tissue gives up, the game is over.

Unlike skeletal muscle and smooth muscle, contraction of the heart muscle is *autonomous,* which means it occurs without stimulation by a nerve. In between contractions, the fibers relax completely (see Chapter 9).

# Smooth muscle

Smooth muscle tissue is found in the walls of organs and structures of many organ systems, including the digestive system, the urinary system, the respiratory system, the cardiovascular system, and the reproductive system. Smooth muscle tissue is fundamentally different from skeletal muscle tissue and cardiac muscle tissue in terms of cell structure and physiological function. However, smooth muscle sarcomeres are similar.

Smooth muscle fibers (cells) are *fusiform* (thick in the middle and tapered at the ends) and arranged to form sheets of tissue. Smooth muscle cells aren't striated. However, smooth muscle contractions utilize the same sliding filament mechanism as skeletal muscle cells (see the next section for more).

Smooth muscle contraction is typically slow, strong, and enduring. Smooth muscle can hold a contraction longer than skeletal muscle. In fact, some smooth muscles, notably the sphincters, are in a constant state of contraction. Childbirth is among the few occasions in life when humans (some humans, anyway) consciously experience smooth muscle contraction (although they don't consciously control it).

# Getting a Grip on the Sliding Filament

A muscle contracts when all the sarcomeres in all the myofibrils in all the fibers (cells) contract all together. The *sliding filament model* describes the fine points of how this happens.

The key to the sliding filament model is the distinctive shapes of the protein molecules *myosin* and *actin,* and their partial overlap in the sarcomere. The special chemistry of ATP supplies the energy for the filaments' movement. The following sections explain how sarcomeres create muscle contraction.

**REMEMBER**

The *sarcomere* is the functional unit within the *myofibril.* Sarcomeres line up end to end along the myofibril.

## Assembling a sarcomere

The sarcomere is composed of *thick filaments* and *thin filaments.* The thick filaments are molecules of the protein *myosin,* which is dense and rubbery. The thin filaments are made up of two strands of the lighter (less dense) protein *actin,* which is springy. The thin and thick filaments line up together in an orderly way to form a sarcomere. One end of a thin filament touches and adjoins the end of another thin filament forming the Z line that runs perpendicular to the filament axis. The sarcomere begins at one Z line and ends at the next Z line. The thick filaments line up precisely between the thin filaments. Sarcomeres and Z lines are shown in Figures 6-2 and 6-3.

The two types of filaments overlap only partially when the sarcomere is at rest. The partial overlap gives skeletal and cardiac muscle cells their *striations:* where thick and thin filaments overlap, the tissue appears dark (dark band); where only thin filaments are present, the tissue appears lighter (light band).

The myosin filaments have numerous club-shaped heads that point away from its center (toward the two Z lines). These heads rest nearly touching the *myosin binding sites* on the actin. Why, then, do they not link up? Wrapping around the actin is the *troponin-tropomyosin complex.* When a muscle is at rest, this protein covers up the binding sites. Until the myosin heads can link up with the actin, the fiber cannot contract. (See Figure 6-3a for a sarcomere at rest.)

a — Actin — Z-line — Myosin

b — Myosin head — Binding site — Actin

Grab ⟶ Pull in ⟶ Let go and cock back

c

**FIGURE 6-3:**
Sarcomere structure and shortening: a) sarcomere before contraction; b) close-up view of a power stroke; c) contracted sarcomere showing Z lines closer together.

*Illustration by Kathryn Born, MA*

# Telling the fiber to contract

Motor neurons provide the stimulus for a muscle fiber to contract. From the end of its axon, a motor neuron releases *acetylcholine* (ACh), which binds to the muscle cell in a designated area called the *motor end plate.* This triggers the fiber to generate an impulse, which spreads through the sarcolemma (for more on nervous communication and impulse, see Chapter 7). The impulse is spread deep into the cell by tunnels called *T-tubules.* When the impulse reaches a *sarcoplasmic reticulum* (SR), it is triggered to release the calcium ions stored there. The ions move to the sarcomeres, allowing contraction to begin. Triggering relaxation is a passive process. The neuron simply stops releasing ACh. Muscle impulse stops and the calcium ions return to the SR.

## Contracting and releasing the sarcomere

After calcium is released from the SR, it attaches to the troponin. In doing so, it is pulled away from the actin — like pulling a push pin out of a bulletin board. Because troponin is still tightly bound to the tropomyosin, the whole protein complex slides around the actin, exposing the binding sites. Now the myosin heads can link up with the actin forming *cross bridges,* all without the use of energy. However, to actually cause a contraction, we must shorten the sarcomere.

When the cross bridges form, the myosin heads immediately bend, pulling the ends of the sarcomere (Z lines) closer together. ATP will then bind to the myosin, causing it to let go of the actin. The energy in ATP causes the heads to cock back into their original position. They will then form a new cross bridge except it is now further down the actin (similar to how a ratchet wrench works). These *power strokes* continue — grab, pull in, let go, cock back — as long as the binding sites are uncovered and ATP is present. Refer to Figure 6-3 to get a picture of this process.

The sarcomere shortens as the filaments slide past each other. Because the myofibrils are made of sarcomeres that share Z lines (refer to Figure 6-2), and this shortening is occurring in all of them, the two ends of the myofibril are pulled noticeably closer together. Because a muscle fiber is packed full of myofibrils, it is shortened as its ends are pulled together, too. A skeletal muscle is made of numerous fibers bundled in parallel to each other so its ends are pulled together as well. The muscle contracts, pulling whatever it's attached to closer together, all with the force generated by the overlapping of these microscopic filaments (but multiplied thousands of times over).

REMEMBER

The filaments (actin and myosin) do not shrink. They maintain their length at all times. The myosin filaments also do not move. Actin is pulled toward the center, progressively increasing the amount of overlap with the myosin (refer to Figure 6-3c).

In order to relax, the fiber simply stops being told to contract (the motor neuron halts its stimulation). This causes the SR to call back the calcium ions like a mother calling her kids back home for dinner. They happily oblige, letting go of the troponin, which will pin back into the actin. This slides the tropomyosin over the top of the binding sites again, knocking off any of the myosin heads that were attached (breaking the cross bridges). The actin will then slowly slide back restoring the length of the sarcomere to its original position.

## THE LAST CONTRACTION

The time comes for every animal — humans included — to die. The fact that every animal gets cold and stiff tells others when that time has come. Do you know why? The cells no longer make ATP.

At the moment of death, the lungs stop filling with oxygen, the heart stops pumping blood through the body, and the brain stops sending signals. The cells — without incoming oxygen, nutrients, or stimulus from the brain — cease performing their metabolic reactions. So ATP can no longer be produced.

Without ATP flooding the myofibrils, contractions can't occur, but neither can the last step of muscle contraction, the step that allows the muscles to relax. In order for a myofibril to relax, ATP must hook onto myosin and dissolve the actin-myosin cross-bridges. But when ATP is unavailable to generate a subsequent contraction, the last contraction becomes permanent, and the corpse stiffens. *Rigor mortis,* which means *rigidity of death,* occurs in every muscle throughout the body. And remember that movement of muscles generates heat, so when the muscles stop their physiological reactions and warm blood stops flowing through the blood vessels, the corpse gets cold.

# Naming the Skeletal Muscles

Get ready, because we're about to tell you the muscle names from head to toe — literally. Check out the "Muscular System" color plate in the center of the book as you go through this section.

To name muscles, anatomists had to come up with a set of rules to follow so the names would make sense. They chose to focus on certain characteristics to derive each muscle's Latin name. As you go through the following sections, refer to Chapter 1 if necessary for the names of the body regions. Examples of characteristics in muscle names are given in Table 6-2.

**TABLE 6-2**     ## Characteristics in Muscle Names

| Characteristics | Examples |
|---|---|
| Muscle size | The largest muscle in the buttocks is the *gluteus maximus* (*maximus* means *large* in Latin); a smaller muscle in the buttocks is the *gluteus minimus* (*minimus* means *small* in Latin). |
| Muscle location | The *frontalis muscle* lies on top of the skull's frontal bone. |

*(continued)*

**TABLE 6-2** *(continued)*

| Characteristics | Examples |
|---|---|
| Muscle shape | The *deltoid muscle,* shaped like a triangle, comes from *delta* — the Greek alphabet's fourth letter, which is also shaped like a triangle. |
| Muscle action | The *extensor digitorum* is a muscle that extends the fingers or *digits.* |
| Number of muscle attachments | The *biceps brachii* attaches to bone in two locations, whereas the *triceps brachii* attaches to bone in three locations. |
| Muscle fiber direction | The *rectus abdominis muscle* runs vertically along your abdomen (*rectus* means *straight* in Latin). |

## Starting at the top

Your head contains muscles that perform three basic functions: chewing, making facial expressions, and moving your neck. Ear wiggling falls into this category, too.

To chew, you use the muscles of *mastication* (a big, fancy word that means "chewing"). The *masseter,* a muscle that runs from the *zygomatic bone* (your cheekbone) to the *mandible* (your lower jaw), is the prime mover for mastication, so its name is based on its action (*mas*seter, *mas*tication). The fan-shaped *temporalis muscle* works with the masseter to allow you to close your jaw. It lies on top of the skull's temporal bone, so its name is based on its location. Figure 6-4 shows the muscles of the head and neck.

To smile, frown, or make a funny face, you use several muscles. The *frontalis muscle* along with a tiny muscle called the *corrugator supercilii* raises your eyebrows and gives you a worried or angry look when wrinkling your brow. (Think of the appearance of corrugated cardboard, and then feel the skin between your eyebrows when you wrinkle your brow.) The *orbicularis oculi muscle* surrounds the eye (the word *orbit,* as in *orbicularis,* means "to encircle"; *oculi* refers to the eye). This muscle allows you to blink your eyes and close your eyelids, but it also gives you those little crow's feet at the corners of your eyes. The *orbicularis oris* surrounds the mouth. (*Or* refers to mouth, as in "oral.") You use this muscle to pucker up for a kiss. Figure 6-5 shows the facial muscles.

If you play the trumpet or another instrument that requires you to blow out, you're well aware of what your *buccinator muscle* does. This muscle is in your cheek. (*Bucc* means "cheek," as in the word *buccal,* which refers to the cheek area.) It allows you to whistle and also helps keep food in contact with your teeth as you chew. Remember that your zygomatic is your cheekbone? Well, the *zygomaticus muscle* is a branched muscle that runs from your cheekbone to the corners of your mouth. This muscle pulls your mouth up into a smile when the mood strikes you.

**FIGURE 6-4:**
The muscles
of the head
and neck.

Cranial aponeurosis

Temporalis

Frontalis

Orbicularis oculi

Levator labii superioris
alaque nasi

Levator labii superioris

Zygomaticus major
and minor

Orbicularis oris

Depressor anguli oris

Risorius

Masseter

Sternocleidomastoid

Occipitalis

Buccinator

Splenius capitus

Levator scapulae

Trapezius

Middle scalene

Omonyoid

Anterior scalene

*Illustration by Kathryn Born, MA*

**FIGURE 6-5:**
The muscles of
the face.

Frontalis

Corrugator supercilii

Orbicularis oculi

Levator labii superioris

Zygomaticus major
and minor

Masseter

Depressor anguli oris

Mentalis

Sternocleidomastoid

Nasalis

Levator labii superioris alaque nasi

Risorius

Orbicularis oris

Trapezius

Note: Sternocleidomastoid has two parts.

*Illustration by Kathryn Born, MA*

When you want to nod *yes, no,* or tilt your head into a *maybe so,* your neck muscles come into play. You have two *sternocleidomastoid muscles,* one on each side of your neck. We know this is a long name, but the name reflects the locations of its attachments: the sternum, clavicle, and mastoid process of the skull's temporal bone. When both sternocleidomastoid muscles contract, you can bring your head down toward your chest and flex your neck. When you turn your head to the side, one sternocleidomastoid muscle contracts — the one on the opposite side of the direction your head is turned. So if you turn your head to the left, your right sternocleidomastoid muscle contracts, and vice versa. If you lean your head back to look up at the sky or to shrug your shoulders, your *trapezius muscle* allows you to do so.

The trapezius is an antagonist to the sternocleidomastoid muscle. If you remember basic geometry, you can see that the trapezius is shaped like a trapezoid. It runs from the base of your skull to your *thoracic vertebrae* and connects to your scapula (shoulder blades). Therefore, the trapezius and sternocleidomastoid muscles connect your head to your torso and provide a nice segue to the next section. The trapezius is shown in Figure 6-7.

## Twisting the torso

**REMEMBER**

The torso muscles have important functions. They not only give support to your body but also connect to your limbs to allow movement, allow you to inhale and exhale, and protect your internal organs. In this section, we cover the muscles that run along the front of you (called your anterior or ventral side) and then cover the muscles of your back (your posterior or dorsal side).

In your chest (see Figure 6-6), your *pectoralis major muscles* connect your torso at the sternum and collarbones (clavicles) to your upper limbs at the humerus bone in the upper arm. Your "pecs" also help to protect your ribs, heart, and lungs. You can feel your pectoralis major muscle working when you move your arm across your chest. Also in your chest are the muscles between and around the ribs. The *internal intercostal muscles* help to raise and lower your rib cage as you breathe. However, the torso's largest muscles are the abdominal muscles.

**TIP**

The *abdominal muscles* really form the center of your body. If the abdominal muscles are weak, the back is weak because the abdominal muscles help to flex the vertebral column. So if the vertebral column doesn't flex easily, the muscles attached to it can become strained and weak. And the muscles of the abdomen and back join to the upper and lower limbs. Therefore, if the abdomen and back are weak, the limbs can have problems.

Pectoralis minor

Pectoralis major

Rectus abdominis

Cut edge of external oblique

Internal oblique

External oblique

Cut edge of aponeurosis
of internal oblique

Transversus abdominis

K. BORN

**FIGURE 6-6:**
Anterior muscles
of the chest and
abdomen.

*Illustration by Kathryn Born, MA*

The muscles of the abdomen are thin, but the fact that these muscle fibers run in different directions increases their strength. This woven effect makes the tissues much stronger than they would be if they all went in the same direction. Think about how a child connects building blocks. Laying a top layer of blocks perpendicular to the blocks underneath helps the structure stay together, which is similar to how the abdominal muscle tissues provide strength and stability.

The "six-pack" muscle of the abdomen, the *rectus abdominis,* forms the front layer of the abdominal muscles, and it runs from the pubic bone up to the ribs and sternum. The function of the rectus abdominis muscle is to hold in the organs of the abdomino-pelvic cavity and allow the vertebral column to flex.

Other layers of abdominal muscles also help to hold in your organs on the side of your abdomen and provide strength to your body's core. The *external oblique muscles* attach to the eight lower ribs and run downward toward the middle of your body (slanting toward the pelvis). The *internal oblique muscles* lie underneath the external oblique (makes sense, eh?) at right angles to the external oblique muscles. The internal oblique muscles extend from the top of the hip at the iliac crest to the lower ribs.

Together, the external and internal oblique muscles form an *X*, essentially strapping together the abdomen. The abdomen's deepest muscle, the *transversus abdominis,* runs horizontally across the abdomen; its function is to tighten the abdominal wall, push the diaphragm upward to help with breathing, and help the body bend forward. The transversus abdominis is connected to the lower ribs and lumbar vertebrae and wraps around to the pubic crest and *linea alba.* The linea alba ("white line") is a band of connective tissue that runs vertically down the front of the abdomen from the xiphoid process at the bottom of the sternum to the *pubic symphysis* (the strip of connective tissue that joins the hip bones).

The muscles in your back (refer to Figure 6-7) serve to provide strength, join your torso to your upper and lower limbs, and protect organs that lie toward the back of your trunk (such as your kidneys). The *deltoid muscle* joins the shoulder to the collarbone, scapula, and humerus. This muscle is shaped like a triangle (think of the Greek letter delta: Δ). The deltoid muscle helps you raise your arm up to the side (that is, laterally). The *latissimus dorsi muscle* is a wide muscle that's also shaped like a triangle. It originates at the lower part of the spine (thoracic and lumbar vertebrae) and runs upward on a slant to the humerus. Your "lats" allow you to move your arm down if you have it raised, and also to reach, such as when you're climbing or swimming.

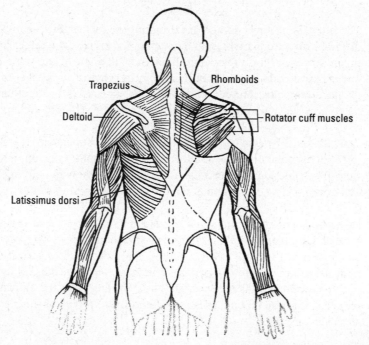

**FIGURE 6-7:**
Posterior view of the neck and torso muscles.

*Illustration by Kathryn Born, MA*

# Spreading your wings

Your upper limbs have a wide range of motions. Obviously, your upper limbs are connected to your torso. One of the muscles that provides that connection, the *serratus anterior,* is below your armpit (the anatomic term for armpit is *axilla*) and on the side of your chest. The serratus anterior muscle connects to the scapula and the upper ribs. You use this muscle when you push something or raise your arm higher than horizontal. Its action pulls the scapula downward and forward.

Although the *biceps brachii* and *triceps brachii* are muscles located in the top (anterior) part of your upper arm, their actions allow your forearm (lower arm) to move. Figure 6-8a shows an anterior view of the upper limb. The name *biceps* refers to this muscle's two origins (points of attachment); it attaches to the scapula in two places. From there, it runs to the radius of the forearm (its point of insertion). The triceps brachii is the only muscle that runs along the back (posterior) side of the upper arm. Figure 6-8b shows a posterior view of the upper limb. The name *triceps* refers to the fact that it has three attachments: one on the scapula and two on the humerus. It runs to the ulna of the forearm. You can feel this muscle in motion when you push or punch. Other muscles of the arm include the *brachioradialis,* which helps you flex your arm at the elbow, and the *supinator,* which rotates your arm from a palm-down position to a palm-up position (remember the word supinator has the word "up" in it).

**FIGURE 6-8:**
The muscles of the upper limb: anterior (a) and posterior (b).

*Illustration by Kathryn Born, MA*

Your forearm contains muscles that control the fine movements of your fingers. When you type or play the piano, you're using your *extensor digitorum* and *flexor digitorum muscles* to raise and lower your fingers onto the keyboard and move them to the different rows of keys. As you lift your hands off the keyboard, the muscles of your wrist kick into gear. The *flexor carpi radialis* (attached to the radius bone) and *flexor carpi ulnaris* (attached to the ulna bone) allow your wrist to flex forward or downward. The *extensor carpi radialis longus* (which passes by the carpal bones), the *extensor carpi radialis brevis*, and the *extensor carpi ulnaris* allow your wrist to extend; that is, bend upward.

The muscles in your upper arm move your forearm. The muscles of your forearm move your wrists, hands, and fingers. There are no muscles in your fingers — only the tendons that connect to those bones.

## Getting a leg up

Your lower limbs are connected to your buttocks, and your buttocks are connected to your hips. The *iliopsoas* connects your lower limb to your torso and consists of two smaller muscles: the *psoas major*, which joins the thigh to the vertebral column, and the *iliacus*, which joins the hipbone's ilius to the thigh's femur bone. Originating on the iliac spine of the hip and joining to the inside surface of the tibia (a bone in your shin), the *sartorius muscle* is a long, thin muscle that runs from the hip to the inside of the knee (see Figure 6-9a). These muscles stabilize your lower limbs and provide strength for them to support your body's weight and to balance your body against the pressure of gravity.

## THUMBS UP!

A key feature of all primates is a *prehensile thumb;* that is, a thumb that's adapted for grasping objects. Many animals have digitlike structures, but only primates can grasp things with their hands. And the only way to grasp things is to have a thumb.

Imagine having webbing between your four fingers so that you couldn't spread them apart; you wouldn't be able to pick things up. That's why animals such as dogs, cats, and birds hold things in their mouth (or beak). But primates — apes, monkeys, and humans — can easily grasp things between their thumb and fingers. However, of those primates, only humans have an *opposable thumb* (one that can touch each of the other fingers; the thumb can be "opposite" from each finger).

Because of the ability to oppose the thumb to each finger, the muscles in your digits are capable of performing minute movements. As you touch your thumb to your pinky finger, your palm becomes arched, which only happens in humans because of short bones in the pinky and an opposable thumb.

Some muscles in the lower limb allow the thigh to move in a variety of positions. The buttocks muscles allow you to straighten your lower limb at the hip and extend your thigh when you walk, climb, or jump. The *gluteus maximus* — the largest muscle in the buttocks — is the largest muscle in the body (see Figure 6-9b). The gluteus maximus is antagonistic to the stabilizing *iliopsoas muscle*, which flexes your thigh. The *gluteus medius muscle*, which lies behind the gluteus maximus, allows you to raise your leg to the side so that you can form a 90-degree angle with your two legs (this action is *abduction* of the thigh). Several muscles serve as *adductors;* that is, they move an abducted thigh back toward the midline. These muscles include the *pectineus* and *adductor longus*, which become injured when you "pull a groin muscle," as well as the *adductor magnus* and *gracilis*, which run along the inside of your thigh.

**FIGURE 6-9:**
The muscles of
the lower limb:
anterior (a) and
posterior (b).

*Illustration by Kathryn Born, MA*

Other muscles in the thigh serve to move the lower leg. Along your thigh's front and lateral side, four muscles work together to allow you to kick. These four muscles — the *rectus femoris, vastus lateralis, vastus medialis,* and *vastus intermedius* — are better known as the *quadriceps (quadriceps femoris). Quad* means "four," as in *quadrilateral* or *quadrant.* Refer to Figure 6-9a.

The *hamstrings* are a group of muscles that are antagonistic to the quadriceps. The hamstrings — the *biceps femoris, semimembranosus,* and *semitendinosus* — run down the back of the thigh (refer to Figure 6-9b) and allow you to flex your lower leg and extend your hip. They originate on the ischium of the hipbone and join (insert at) to the tibia of the lower leg. You can feel the tendons of your hamstring muscles behind your knee.

Your lower leg's shin and calf muscles move your ankle and foot. The *gastrocnemius,* better known as the "calf muscle," begins (originates) at the femur (thigh bone) and joins (inserts at) the *Achilles tendon* that runs behind your heel. You can feel your gastrocnemius muscle contracting when you stand on your toes. The antagonist of the gastrocnemius, the *tibialis anterior,* starts on the surface of the tibia (shinbone), runs along the shin, and connects to your ankle's metatarsal bones. You can feel this muscle contract when you raise your toes and keep your heel on the floor. The *fibular longus* and *fibular brevis (brevis* meaning "short," as in *brevity)* run along the outside of the lower leg and join the fibula to the ankle bones. In doing so, the fibular muscles help to move the foot. The *extensor digitorum longus* and the *flexor digitorum longus* muscles join the tibia to the feet and allow you to extend and flex your toes, respectively, like your fingers.

# Pathophysiology of the Muscular System

The body has so much skeletal muscle tissue, and it performs so many functions that wear and tear is normal and expected, not really "pathophysiological." Sore or strained muscles need only time to repair themselves and even become better than ever. However, many serious conditions can affect skeletal muscle and leave the sufferer disabled, in considerable pain, and even with a much-shortened life span. The following sections give you an overview of conditions that can affect skeletal muscles.

# Muscular dystrophy

The muscular dystrophies (MD) are a group of more than 30 diseases characterized by progressive weakness and degeneration of the skeletal muscles. All are inherited, each in its own pattern. The disorders differ in terms of the distribution and extent of muscle weakness, age of onset of symptoms, and rate of progression. The prognosis for people with MD varies according to the disorder's type and progression. Some cases may be mild and progress very slowly over a normal life span, while others produce severe muscle weakness and functional disability beginning at a young age. Some children with MD die in infancy, while others live into adulthood with only moderate disability.

The most common form is *Duchenne muscular dystrophy* (DMD). Since the mutation is on the X chromosome, it is far more common in males. Females receive two copies of the X (so they have a chance for a properly functioning gene on the second copy) and because the symptoms are so severe, the odds of a male passing on his mutated chromosome are very low.

The symptoms of DMD become evident usually before a child is 3 years old. The muscles slowly weaken, shorten, and degenerate. Fat and connective tissue replace normal muscle tissue, thus causing problems in the heart and lungs. DMD patients are often in a wheelchair by about age 12 and usually die in their teens. Some female carriers of DMD gene do experience symptoms, but they are much milder.

*Myotonic muscular dystrophy* affects males and females, and the onset of symptoms can occur at any age. Progressive muscle weakness and stiffness is usually evident first in the muscles of the face and neck. Turning the head becomes difficult. Eventually, those afflicted have problems with actions such as swallowing, because their muscles don't relax after contractions. Then, the arms and legs become affected. Eventually, the patient may require a wheelchair or even be confined to bed.

There's no specific treatment to stop or reverse any form of MD. Treatment to manage symptoms may include drug therapy, physical therapy, respiratory therapy, speech therapy, orthopedic appliances used for support, and corrective orthopedic surgery. Some patients may need assisted ventilation to treat respiratory muscle weakness and a pacemaker for cardiac abnormalities.

# Muscle spasms

A *muscle spasm* is a sudden, strong, involuntary contraction, sometimes causing severe pain. Any muscle can go into spasm, and the effects vary according to the location and nerves that are nearby, but most often, you feel it as a cramp. Common causes of muscle spasms include overuse, insufficient stretching, and dehydration. Muscle spasms are a common cause of back and neck pain. The *gastrocnemius* (calf muscle) is a common place for sudden cramps to occur — the dreaded charley horse.

Not all spasms are painful. Hiccups, which are the result of spasm in the diaphragm, aren't usually painful — annoying, but not painful. The same with facial tics, such as when your eyelid twitches when you're stressed.

## Fibromyalgia

*Fibromyalgia*, a chronic pain condition, isn't exactly a muscle disease, but severe and widespread muscle pain is a major symptom of this mysterious condition. Strictly speaking, fibromyalgia isn't a disease at all but a *syndrome* — a group of symptoms that appear to be closely related, although the underlying cause may not be known. Recent research has suggested a genetic component. The disorder is often seen in families, among siblings or mothers and their children.

The muscle pain can be more or less severe, more or less chronic, and more or less debilitating. Fibromyalgia patients often have more of a neurotransmitter called *substance P* in their spinal fluid, which is believed by some clinicians to alter the affected person's perception of pain. A sensation that someone else may not even notice can be experienced by the fibromyalgia sufferer as excruciating. Some drug treatments are available, and some pain management techniques as well as stress reduction are helpful for some patients. Fibromyalgia is an active area of clinical research.

# 3
# Talking to Yourself

Look at the anatomical structures of the nervous and endocrine systems.

Consider consciousness and control in the nervous system.

Understand impulse and cell-to-cell communication.

Make sense of the major hormones and their functions.

IN THIS CHAPTER

» Summarizing the functions of the nervous system

» Taking a detailed look at the CNS, the brain, and the PNS

» Getting down to a cellular level — neurons and neuroglial cells

» Relaying impulses through a cell and across a synapse

» Receiving impulses: your five senses

» Noting a few nervous system disorders

Chapter **7**

# The Nervous System: Your Body's Circuit Board

An organism's awareness of itself and of its environment depends on communication between one part of its body and another. In biology, such internal messaging is accomplished by several different mechanisms. Humans, like all mammals, use mechanisms involving chemistry and mechanisms involving electricity. We discuss the chemical messaging system (hormones) in Chapter 8. This chapter is devoted to the body's electrical messaging system.

The nervous system is the body's electrical communications network. It generates and transmits information throughout the body in the form of electrical impulses. An electrical charge creates electrical energy, which has two important characteristics: It moves in distinct "packets," called *impulses*, and it moves very quickly.

The nervous system's structures reach into every organ and participate one way or another in nearly every physiological reaction. Perceiving the beauty of a flying bird and digesting your breakfast can happen simultaneously, and each action is dependent on the nervous system.

The human nervous system is, in fact, the single most distinctive feature of the human species. In particular, the human brain, an organ of the nervous system, functions differently from the brain of any other species. Although just how this system contributes to human nature (consciousness, for example) is being actively researched, the goal here is understanding the mechanics.

# Integrating the Input with the Output

The nervous system has just three jobs to do, and these jobs overlap.

>> **Sensory input:** Specialized neurons called *sensory receptors* collect information from the entire body, create an impulse, and transmit the impulse to either the spinal cord or brain stem, and then to the brain. (See the "Making Sense of Your Senses" section later in this chapter.)

>> **Integration:** The *central nervous system* (CNS) makes sense of the input it receives from all around the body. (See the "Central nervous system" section later in the chapter.)

>> **Motor output:** In response to the integration of the sensory input, impulses are sent out through the *peripheral nervous system* (PNS) to muscles, glands, and other organs capable of the appropriate response. (See the "Peripheral nervous system" section later in the chapter.)

REMEMBER

The PNS is the route of communication between the brain and spinal cord — the central nervous system (CNS) — and the rest of the body. The sensory information coming in does so via the PNS just as the outgoing messages do. When discussing the PNS, important distinctions are most relevant to the motor output, but don't forget that the classification does include sensory input.

## Nervous tissues

Nervous tissue is made up primarily of two categories of cells — neurons and neuroglial cells. The structure and distribution of the cells differ throughout the different specialized tissues of the nervous system.

# Neurons

A *neuron* is an individual cell and is the basic unit of the nervous system. Neurons are highly specialized for the initiation and transmission of electrical signals (impulses). The neuron is able to, in an instant, receive stimuli from many other cells, process this incoming information, and "decide" whether to generate its own signal to be passed on to other neurons, muscles, or gland cells.

Here are the three types of neurons:

>> **Sensory neurons:** Also called *afferent neurons* (*afferent* means "moving toward"), these neurons respond to sensory stimuli (touch, sound, light, and so on), passing the impulses ultimately to the spinal cord and brain.

>> **Motor neurons:** Also called *efferent neurons* (*efferent* means "moving away"), these transmit impulses from the brain and spinal cord to effector organs (muscles and glands), triggering responses from these organs (muscle contraction or release of the gland's product).

>> **Interneurons:** Also called *association neurons,* these connect neurons to other neurons within the same region of the brain or spinal cord.

Neurons in different parts of the nervous system perform diverse functions and therefore vary in shape, size, and electrochemical properties. However, neurons have a special cellular anatomy adapted to the quick transmission of an electrical charge. See Figure 7-1. All neurons have the same three parts, all enclosed within their cell membrane:

>> **Cell body:** The body of a neuron is similar to a generic cell. It contains the nucleus, mitochondria, and other organelles.

>> **Dendrites:** The *dendrites* are extensions that branch from one end of the cell body. They receive information from other neurons and transmit stimulus toward the cell body.

>> **Axon:** Each neuron has a single cable-like projection called an *axon.* Extending many times (tens, hundreds, or even tens of thousands of times) the cell body's diameter in length, the axon carries the impulses away from the cell body and toward the next neuron in the chain. (Think of electrical transmission wires.)

**TIP**

*Axon* means "away." You can remember this because they both start with *a.*

Fully differentiated neurons don't typically divide and may live for years, or even the whole lifetime of an organism. The creation of new neurons is an active area of research.

The longest neuron in your body runs from the tip of your big toe to the base of your spinal cord. That's a single cell around 3 feet long!

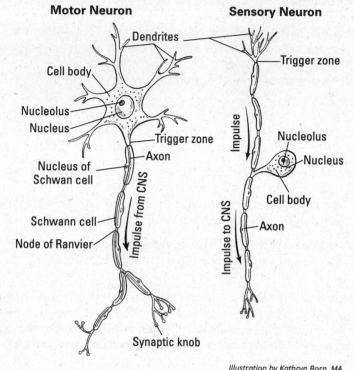

**FIGURE 7-1:**
Motor neuron (a) and sensory neuron (b), structure and path of impulses.

## Neuroglial cells

The numerous cells of nervous tissue that aren't neurons are collectively called *glial cells* (or *neuroglia*). These cells can communicate with each other, but they lack axons and dendrites and do not generate impulses. Though the ratio varies throughout the brain, glial cells outnumber neurons by at least 3:1.

Until recently, most neuroscientists thought glial cells were merely supporting structures, effectively gluing the neurons together (*glia* is Greek for "glue"). However, with new imaging technology and *in vitro* (outside of the body) lab techniques, we've learned that these cells do far more than that. Table 7-1 outlines these functions.

**TABLE 7-1** | **Glial Cell Functions**

| Cell | Location | Functions |
|------|----------|-----------|
| Astrocyte | CNS | Regulates chemicals in the synapse; creates new neuron connections |
| Ependymal cell | CNS | Makes cerebrospinal fluid (CSF) |
| Microglia | CNS | Provides immune protection by phagocytosis |
| Oligodendrocyte | CNS | Forms myelin sheath to speed impulse transmission |
| Schwann cell | PNS | Speeds impulse transmission; promotes axon regeneration |

## Nerves

A *nerve* is a bundle of peripheral axons. An individual axon plus its myelin sheath is called a *nerve fiber*. Nerves provide a common pathway for the electrochemical nerve impulses that are transmitted along each of the axons. Nerves are found only in the peripheral nervous system. Nerve fibers can be of two types: *motor*, which send impulses away from the CNS, or *sensory*, which send impulses toward the CNS.

## Ganglia and plexuses

A *ganglion* (plural, *ganglia*) is an aggregation of neuron cell bodies. Ganglia provide relay points among the body's neurological structures, especially at the spinal cord, serving as the junction between the CNS and the PNS.

Ganglia may be connected to form a *chain*. For example, the sympathetic nervous system contains a chain of ganglia referred to as the *paravertebral ganglia* or the *sympathetic chain of ganglia* that spans the length of the spinal cord.

*Plexus* is a general term for a network of anatomical structures, such as lymphatic vessels, nerves, or veins. (The term comes from the Latin "plectere" meaning "to braid.") A neural plexus is a network of intersecting nerves. The *solar plexus* serves the internal organs. The *cervical plexus* serves the head, neck, and shoulders. The *brachial plexus* serves the chest, shoulders, arms, and hands. The *lumbar*, *sacral*, and *coccygeal plexuses* serve the lower body.

# Integrated Networks

The nervous system comprises two physically separate but functionally integrated networks of nervous tissue. Working together, these networks perceive and respond to internal and external stimuli to maintain homeostasis and simultaneously move the genetic development program forward. The following sections take a closer look at the central and peripheral nervous systems. Check out the "Nervous System" color plate in the middle of the book for a detailed look at the anatomy.

# Central nervous system

The *central nervous system* (CNS), which consists of the brain and spinal cord, is the largest part of the nervous system. It integrates the information it receives from the sensory receptors and coordinates the activity of all parts of the body.

Both the brain and spinal cord are masses of neural tissue protected within bony structures (the skull and the vertebral column, respectively) and layers of membranes and specialized fluids, reflecting their prime importance to the continuation of the organism's life.

The brain and spinal cord are made up mainly of two types of tissue, called *gray matter* and *white matter.* Gray matter consists of unmyelinated neurons, neuron cell bodies, and neuroglial cells. White matter is made up of neuroglial cells and the myelinated axons extending from the neuron cell bodies in the gray matter. (See the "Nervous Tissues" section earlier in the chapter for more on neurons and neuroglial cells.) The myelin has a high lipid content, which results in white matter's white color.

In the brain, the gray matter forms a thin layer on the outside (the *cortex*). The white matter is beneath and makes up the brain's big data lines, carrying information around the brain. In the spinal cord, the tissue is arranged in a long cylinder; the gray matter forms the inner layer, and the white matter forms the outer layer.

The spinal cord extends from the bottom of the brain stem down the vertebral column within a cylindrical tubular opening created by the vertebrae and three tough membranes with cushioning fluid between them (see Figure 7-2).

The three membranes that surround the spinal cord, collectively called the *meninges,* continue up to encase the brain. The outermost layer is the *dura mater,* which is followed by the *arachnoid mater.* Between these two layers flows a fluid much like interstitial fluid. The innermost membrane, the *pia mater,* contacts the nervous tissue. Between the arachnoid and pia flows a fluid unique to the CNS called *cerebrospinal fluid* (CSF).

Spinous process

Spinal ganglion

Spinal cord

Meninges: Dura mater
Arachnoid mater

Dorsal ramus
of spinal nerve

Dorsal root

Ventral ramus
of spinal nerve

Ventral root

Subarachnoid
space

Sympathetic
ganglion

Gray and
white rami

Fat in epidural
space

Vertebral body
(of a throracic
verebra)

K. BORN

**FIGURE 7-2:**
Cross section of
the spinal cord.

*Illustration by Kathryn Born, MA*

# Peripheral nervous system

The *peripheral nervous system* (PNS) consists of the nerves and ganglia outside of the brain and spinal cord. Unlike the CNS, the PNS isn't protected by bone or by the blood–brain barrier, leaving it exposed to toxins and mechanical injuries. Structures of the PNS include:

>> **Cranial nerves:** Twelve pairs of nerves that emerge directly from the brain and brain stem. Each pair is dedicated to particular functions — some bring information from the sense organs to the brain, others control muscles, but most have both sensory and motor functions. Some are connected to glands or internal organs such as the heart and lungs. For example, the longest of the cranial nerves, called the *vagus nerves,* pass through the neck and chest into the abdomen. They relay sensory impulses from part of the ear, tongue, larynx, and pharynx; relay motor impulses to the vocal cords; and relay motor and secretory impulses to some abdominal and thoracic organs.

>> **Spinal nerves:** Thirty-one pairs of nerves that emerge from the spinal cord. Each contains thousands of *afferent* (sensory) and *efferent* (motor) fibers.

>> **Sensory nerve fibers:** Nerve fibers all over the body that send impulses to the CNS via the cranial nerves and spinal nerves.

» **Motor nerve fibers:** Nerve fibers that connect to muscles and glands and send impulses from the CNS via the cranial and spinal nerves.

The PNS is further divided into the somatic system and the autonomic system.

## Somatic system

The *somatic nervous system* regulates activities that are under conscious control. Its sensory fibers receive impulses from receptors. Its motor fibers transmit impulses from the CNS to the (voluntary) skeletal muscles to coordinate body movements.

## Autonomic system

The motor fibers of the *autonomic system* transmit impulses from the CNS to the glands, heart, and smooth (involuntary) organ muscles. The autonomic system controls internal organ functions that are involuntary and that happen subconsciously, such as breathing, heartbeat, and digestion.

The autonomic system is made up of the following:

» **Sympathetic nervous system:** Nerves originate in the thoracic and lumbar regions of the spinal cord. The sympathetic nervous system responds to stress and is responsible for the increase of your heartbeat and blood pressure, among other physiological changes, along with the sense of excitement you may feel due to the increase of adrenaline in your system. This system is responsible for your "fight or flight" response.

» **Parasympathetic nervous system:** Nerves originate in the brain stem and sacral portion of the spinal cord. The parasympathetic nervous system is evident when you rest or feel relaxed and is responsible for such things as the constriction of the pupil, the slowing of the heart, the dilation of the blood vessels, and the stimulation of the digestive and urinary systems. The parasympathetic nervous system is known as the "housekeeping" system because it maintains your normal functioning when you're not under stress.

» **Enteric nervous system:** This system manages every aspect of digestion, from the esophagus to the stomach to the small intestine to the colon.

# Thinking about Your Brain

The brain is one of the largest organs in the human body, third to the skin and liver. It makes up 3 percent of our body weight but uses about 20 percent of our energy, illustrating its importance. As the task master, the brain manages its workload by compartmentalizing its functions. The different parts of the brain are in constant contact with each other, influencing one another, but they're responsible for different functions.

The brain's major parts are the *cerebrum*, *cerebellum*, *brain stem*, and *diencephalon*. The brain's four connecting, fluid-filled cavities are called *ventricles*. In this section, you find out some details about the parts of your brain and its ventricles. Take a look at Figure 7-3, and refer to it as necessary.

**FIGURE 7-3:** Basic brain anatomy.

*Illustration by Kathryn Born, MA*

# Keeping conscious: Your cerebrum

If you're conscious, you're using your cerebrum. The *cerebrum*, the brain's largest part, controls consciousness among other advanced functions.

The cerebrum is divided into left and right halves, called the *left* and *right cerebral hemispheres*, and each half has four lobes: *frontal, parietal, temporal,* and *occipital*. The names of the lobes come from the skull bones that overlay them (see Chapter 5). Table 7-2 shows you what each lobe controls.

**TABLE 7-2**      **Functions of Lobes within Cerebral Hemispheres**

| Lobe | Functions |
|---|---|
| Frontal lobe | Speech production, concentration, problem solving, planning, and voluntary muscle control |
| Parietal lobe | General interpretation area, understanding speech, ability to use words, and sensations including heat/cold, pressure, touch, and pain |
| Temporal lobe | Interpretation of sensations, remembering visually, remembering through sounds, hearing, and learning |
| Occipital lobe | Vision, recognizing objects visually, and combining images received visually with other senses |

The *cerebral cortex* is the brain's outer layer of gray matter. It covers the entire surface of the cerebrum and overlays the deeper white matter. The brain's elevations are called *gyri* (singular *gyrus*). The shallow grooves that separate the elevations are called *sulci* (singular *sulcus*). Deep grooves in the brain are called *fissures*.

When you look at the top of a brain, you notice a deep groove running down the middle of the cerebrum. This groove is the *longitudinal fissure,* and it incompletely divides the cerebrum into the left and right hemispheres. The *corpus callosum,* located within the brain at the bottom of the longitudinal fissure, contains myelinated fibers that connect the left and right hemispheres.

# Making your moves smooth: The cerebellum

The *cerebellum* lies just below the back half of the cerebrum. A narrow stalk *(vermis)* connects the cerebellum's left and right hemispheres. The outside is gray matter, and the inside is white matter.

The cerebellum coordinates your skeletal muscle movements, making them smooth and graceful instead of stiff and jerky. The cerebellum also maintains normal muscle tone and posture, utilizing sensory information from the eyes, inner ear, and muscles.

# Coming up roses: Your brain stem

The *brain stem* consists of the *midbrain, pons,* and *medulla oblongata.* The medulla oblongata is continuous with the spinal cord after it passes through the hole called the *foramen magnum* at the bottom of the skull (see Chapter 5 for more on the skeletal system).

Inside your brain, just in front of (anterior to) the cerebellum, lie the midbrain and pons. The midbrain serves as a "station" for information that passes between the spinal cord and the cerebrum or between the cerebrum and the cerebellum. Impulses pass through the midbrain, which has centers for reflexes based on vision, hearing, and touch. If you see, hear, or feel something that scares you, alarms you, or hurts you, your midbrain immediately responds by sending out impulses to generate the appropriate type of scream, jump, or exclamation.

*Reflex arcs* sometimes create immediate, unconscious responses. Reflex arcs happen unconsciously whenever you touch something really hot or sharp. Sensory neurons detect pain, temperature, pressure, and the like. If sensory neurons detect something that could harm your body, such as heat that may cause a burn or a sharp object that may puncture the skin, an impulse passes from the receptor in the skin through the sensory neuron to the spinal cord, and then to motor neurons that cause a muscle to contract and pull the body part at risk of injury away from the heat or sharp object.

Reflexes occur so fast that you don't have time to think about how to react. By the time the impulse gets to your brain, the spinal cord has already taken care of the problem! In normal processes of the CNS, impulses travel to the brain for interpretation and production of the proper response. However, using the spinal cord rather than the brain to produce a response, reflex arcs save time and possibly damaging consequences.

If the midbrain is a station for impulses, the pons is the bridge that joins the cerebellum with the cerebrum's left and right hemispheres, allowing the cerebrum to influence the cerebellum. Axon bundles fill the pons and respond quickly to information it receives through the eyes and ears. (See the "Neurons" section earlier in the chapter for more on axons.)

The medulla oblongata, which transitions into the spinal cord, is responsible for several important functions, such as your breathing, the beating of your heart, and the regulation of your blood pressure. The medulla oblongata also contains the axons that send out the signals for coughing, vomiting, sneezing, and swallowing, based on information it receives from the respiratory or digestive systems. And whenever you get those annoying hiccups, blame your medulla oblongata.

## Regulating systems: The diencephalon

Right smack in the middle of the brain, the *hypothalamus* and the *thalamus* form the *diencephalon.* The hypothalamus regulates sleep, hunger, thirst, body temperature, blood pressure, and fluid level to maintain homeostasis. (Flip to Chapter 2 for an overview of homeostasis.)

The thalamus is the gateway to the cerebrum. Whenever a sensory impulse travels from somewhere in your body (except from the nose — sensations of smell are sent directly to the brain by the olfactory nerve), it passes through the thalamus. The thalamus then relays the impulse to the proper location in the cerebral cortex, which then interprets the message. Think of the thalamus as an email server, routing your message through the correct lines.

## Following fluid through the ventricles

Each cerebral hemisphere contains a *lateral ventricle* (the first and second ventricles). The other two ventricles are, believe it or not, the third and fourth ventricles. (Recall that a ventricle is a connecting, fluid-filled cavity.) The *third ventricle* lies just about in the center of your brain; the *fourth ventricle* lies at the top of the brain stem. The *cerebral aqueduct* (also called the *mesencephalic aqueduct*) connects the third and fourth ventricles. From the inferior portion of the fourth ventricle, a narrow channel called the *central canal* continues down into the spinal cord.

TIP

When you think of an aqueduct, you may think of Rome. The Romans built aqueducts as a system for distributing water. Well, in your central nervous system, the ventricles and aqueducts serve as a system to circulate *cerebrospinal fluid* (CSF).

A clear fluid made in the brain, CSF is contained in the brain's four ventricles, the *subarachnoid space* (the space between the arachnoid and the pia mater), and the central canal of the spinal cord. The CSF picks up waste products from the CNS cells and delivers them to the bloodstream for disposal. The CSF also cushions the CNS. Along with your skull and vertebrae, CSF adds a protective layer around your brain and spinal cord.

## THE MOST INTERESTING SYSTEM OF ALL

If you've ever fallen in love, enjoyed sex, stashed away some great memories, or felt enraged (seems like the full cycle of a relationship, doesn't it?), you were using your limbic system. The *limbic system* isn't an anatomical structure but rather a collection of areas in the brain — certain parts of the cerebrum and diencephalon — that are involved in some emotional issues. These areas control your libido, memory, pleasure or pain, and feelings such as happiness, sorrow, fear, affection, and rage. Although these reactions and emotions may not be key to your survival, they do make life interesting.

Perhaps the most important function of CSF is to keep the ions in balance and thus stabilize membrane potentials (more on that in the "Across the neuron" section). CSF circulates from the lateral ventricles to the third ventricle, through the cerebral aqueduct into the fourth ventricle, and then down through the central canal of the spinal cord. From the fourth ventricle, CSF oozes into the subarachnoid space just under the arachnoid membrane, which continuously covers the spinal cord and brain. In the subarachnoid space, CSF can seep through tiny spaces to get to the bloodstream.

In a procedure known as a *spinal tap*, CSF is drawn through a needle from the subarachnoid space. Doctors can test the CSF for the presence of bacteria that cause meningitis or for the presence of proteins that can indicate other diseases, such as Alzheimer's.

## Blood-brain barrier

Blood entering the CNS must pass through the *blood-brain barrier*, which restricts the entry of certain blood molecules into the CNS. The barrier consists of intercellular connections, called *tight junctions*, between the endothelial cells of the capillaries and the processes of the astrocytes that surround them. *Endothelial cells* restrict the diffusion of bacteria, including many common pathogens, thereby protecting the brain from infection. They also block large or hydrophilic molecules from entering the CSF, including some toxins and some drugs. The blood-brain barrier permits the diffusion of small hydrophobic molecules (oxygen, hormones, and carbon dioxide). Other molecules, such as glucose, are carried across the barrier by facilitated diffusion. This works in conjunction with the meninges to tightly control the contents of the fluid that puts it into contact with the nervous tissue of the spinal cord and brain.

# Transmitting the Impulse

To move a message from one part of your environment (whether internal or external) to your brain or spinal cord, an impulse must pass through each neuron and continue on its path. Through a chain of chemical events, the dendrites receive the stimulus, which generates an impulse that travels through the cell to the end of an axon, where a neurotransmitter is released, generating an impulse in the next neuron. The entire impulse passes through a neuron in about seven milliseconds, faster than a lightning strike. We take a look at this lickety-split transmission in detail in the following sections. Refer to Figure 7-4 to visualize this process.

## Across the neuron

When a neuron isn't stimulated (when it's "resting," that is — just sitting with no impulse to transmit), its membrane is *polarized:* The electrical charge on the outside of the membrane is positive while the electrical charge on the inside is negative. The fluid outside the cell contains excess $Na^+$ (sodium ions); the cytoplasm contains excess $K^+$ (potassium ions). This gradient is maintained by *$Na^+/K^+$ pumps* on the membrane. When the neuron is inactive and polarized, it's said to be at its *resting potential*, which is about −70mV.

**TECHNICAL STUFF**

In addition to the $K^+$, negatively charged protein and nucleic acid molecules also inhabit the cell; therefore, the inside is negative relative to the outside.

When a stimulus reaches a dendrite, it open ions channels in the cell membrane. The stimulus could be a neurotransmitter from another neuron or it could be an initiating stimulus on a receptor (for example, temperature increase on a thermoreceptor). When the ion channels open, positive ions (usually $Na^+$) rush in; they're attracted to the negative charge inside the cell. The ions move through the cytoplasm toward the *trigger zone.* In motor neurons, this is where the axon meets the cell body. In sensory neurons, this is where the singular axon meets up with the dendrites (refer to Figure 7-1).

The occurrence of a stimulus does not mean the neuron will generate an impulse. Neurons can only "fire" (create an impulse) or not. For example, the neuron attached to a thermoreceptor can't say "I'm a little warm" versus "I'm burning up!" It either says "Yes, I'm hot" or nothing at all. As a result, the influx of ions must create enough of a voltage change to warrant a response being sent. This is the role of the trigger zone; it determines if the *threshold stimulus* has been met. For most neurons, this is −55mV. If a neuron is stimulated, and positive ions move in increasing the charge to −60mV, then nothing is going to happen. The cell will pump the ions back out to reestablish its resting potential (−70mV). However, the bigger the stimulus (greater force or temperature change, for example), the greater number of ion channels open. Even more positive ions rush in. When the voltage increase hits −55mV at the trigger zone, the axon begins its task of generating impulse.

FIGURE 7-4:
Impulse
transmission:
stimulus of
dendrite (a),
generation and
spread of action
potential (b),
and recovery
wave (c).

③ Ions move
to trigger
zone

Trigger zone

a

① Channels open.

② Ions
move in.

Stimulus

Axon remains at resting
potential (−70 mv).

④ Trigger zone
reaches −55 mv

⑥ Influx of Na⁺ opens the
next set of channels.

b

⑤ Sodium channels open
(1ˢᵗ action potential)

⑧ Continued influx of Na⁺ closes sodium channels
and opens potassium channels.

AP     AP     AP     AP

c

+30 + + + + +

To
synaptic
knob

⑨ K⁺ exits the
cell to repolarize
the axon.

⑦ Action potentials repeat
until synaptic knob reached.

*Illustration by Kathryn Born, MA*

**REMEMBER**

Neurons can't send messages of value to the brain. That information comes from the number of neurons sending the same message.

There are two types of ion channels found along the axon: one for sodium (Na⁺) and the other for potassium (K⁺). Both of these are voltage gated, meaning a specific electrical charge must be reached and then they will pop right open. For the sodium channels, this is −55mV. However, this voltage change is localized so doors only open in the first segment of the axon (see Figure 7-4b). Na⁺ ions rush

CHAPTER 7 **The Nervous System: Your Body's Circuit Board** 157 in, *depolarizing* that segment (making it more positive). Enough ions enter the axon to increase the charge to about 30mV, and because ions can flow down the axon (toward the synaptic knob) the next section quickly reaches −55mV and opens more doors. Each time a set of sodium channels opens on the axon is referred to as an *action potential.* These occur one right after the other until the synaptic knob is reached (see Figure 7-4c), which is the definition of the term *impulse.*

So the impulse spreading down the axon is really several action potentials, each one triggering the next. As such, a neuron can't directly pass impulse to another neuron. When the final action potential reaches the synaptic knob, a slightly different process will trigger the release of neurotransmitters. More on that in the "Across the synapse" section because we're not done with the axon.

We spread the impulse to the end, which is great because that was our goal (and we even did it without the input of energy). However, it'd be nice if we could use that neuron again in the very near future. (*Remember:* We're talking milliseconds here.) Unfortunately, we have a problem. All those Na⁺ ions have depolarized our cell; we have to get them out. Okay, let's open those doors again and let them leave. Nope. As long as the doors are open, Na⁺ will move in because it wasn't just the charge difference attracting them; there's also a concentration gradient (see Chapter 3). This is where the potassium ions come in handy.

When the charge reaches 30mV in an axon segment, the sodium doors are snapped shut and the potassium ones open. K⁺ ions rush out, taking their positive charge with them. This *repolarizes* the membrane. When resting potential (−70mV) is reestablished, the potassium channels close. We still have a problem, though: Our ions are in the wrong place. At this point, the cell will use energy to pump Na⁺ back out and K⁺ back in. Only then will the neuron be capable of firing again. The time it takes for this recovery is known as the neuron's *refractory period.*

**TECHNICAL STUFF**

Many neurons are *myelinated,* having axons wrapped in insulating lipids created by glial cells: oligodendrocytes in the CNS and Schwann cells in the PNS. (The neurons illustrated in Figure 7-1 are both myelinated.) The myelin is said to speed up impulse transmission. What it does is allow the Na⁺ to be shuttled further down the axon before generating the next action potential. This decreases the number of action potentials required to travel the length of the axon, thereby decreasing the time of impulse transmission.

## Across the synapse

Most neurons don't touch. A gap called a *synapse* or *synaptic cleft* separates the axon of one neuron from the dendrite of the next. To traverse this gap, a neuron releases a chemical called a *neurotransmitter* that may or may not cause the next neuron to generate an electrical impulse.

When the impulse reaches the synaptic knob, it again triggers ion channels to open. Only this time, calcium ions rush in. The $Ca^{2+}$ that enter effectively push the packets (vesicles) of neurotransmitters toward the membrane. The neurotransmitters are then released into the synapse where they can bind with any dendrites in the area that have the matching receptor. Figure 7-5 illustrates this process.

**FIGURE 7-5:** Synaptic transmission.

*Illustration by Kathryn Born, MA*

Each type of neurotransmitter has its own type of receptor. Whether the postsynaptic (receiving) neuron is excited or inhibited depends on the neurotransmitter and its effect. For example, if the neurotransmitter is *excitatory*, the $Na^+$ channels open, the neuron membrane becomes depolarized, and the impulse is carried through that neuron. If the neurotransmitter is *inhibitory*, the $K^+$ channels open, the neuron membrane becomes hyperpolarized as the ions exit, and any ongoing impulse is stopped.

# NEUROTRANSMITTERS: FACILITATING THE SIGNAL

*Neurotransmitters* are chemicals that, when released from the end of a neuron's axon, either excite or inhibit an adjacent cell. For example, neurotransmitters are released from the axon endings of motor neurons, where they stimulate or inhibit muscle fibers or glandular cells. Here are some details about some of the better-known neurotransmitters.

- *Acetylcholine* has many functions, notably the stimulation of muscles, including the digestive system's smooth muscles and all skeletal muscles. It's also found in sensory neurons and in the autonomic nervous system, and it plays a part in scheduling REM (dream) sleep.

- *Norepinephrine* (or *noradrenaline*) is released by the adrenal glands, along with its close relative *epinephrine (adrenaline).* It works to enhance the sympathetic nervous system, bringing the nervous system into high alert and increasing the heart rate and blood pressure. It's also important for forming memories.

- *Dopamine,* related to norepinephrine and epinephrine, is strongly associated with reward mechanisms in the brain. If it feels good, dopamine neurons are probably involved. Too little dopamine in the brain's motor areas causes Parkinson's disease, which involves uncontrollable muscle tremors.

- *GABA* (gamma-aminobutyric acid), usually an inhibitory neurotransmitter, acts like a brake to the *excitatory neurotransmitters* that lead to anxiety. People with too little GABA tend to suffer from anxiety disorders, and drugs like Valium work by enhancing the effects of GABA. If GABA is lacking in certain parts of the brain, epilepsy results.

- *Glutamate* is an excitatory relative of GABA. It's the most common neurotransmitter in the central nervous system and is especially important in memory formation. Curiously, glutamate is actually toxic to neurons, and an excess will kill them. ALS, more commonly known as Lou Gehrig's disease, results from excessive glutamate production. Many researchers believe that excess glutamate may also be responsible for quite a variety of nervous system diseases and are looking for ways to minimize its effects.

- *Serotonin* is an inhibitory neurotransmitter that's intimately involved in emotion and mood. Serotonin also plays a role in perception. Serotonin depletion is linked with clinical depression, problems with anger control, obsessive-compulsive disorder, and suicide. It has also been tied to migraines, irritable bowel syndrome, and fibromyalgia.

- *Endorphins* (short for "endogenous morphine") are structurally similar to the *opioids* and have similar functions. They are inhibitory and involved in pain reduction and pleasure. Opioid drugs work by attaching to endorphin receptor sites.

After the neurotransmitter produces its effect (excitation or inhibition), the receptor releases it, and the neurotransmitter goes back into the synapse. In the synapse, one of three things happens to the neurotransmitter. It may

>> Be broken down by an enzyme

>> Be taken back up by the cell that released it (a form of recycling)

>> Simply float away into the surrounding fluid

# Making Sense of Your Senses

Some anatomists consider the sensory system to be part of the peripheral nervous system. Others regard it as a separate system. Either way, there are as many as 21 different senses (depending on how they're grouped). The senses all utilize varieties of five categories of receptor. See Table 7-3 for their description.

**TABLE 7-3**  **Sensory Receptor Types**

| Receptor | Stimulus | Examples |
|---|---|---|
| Chemoreceptor | Chemical change | Smell and taste, blood glucose, pH |
| Nociceptor (pain) | Tissue damage | Widely distributed through skin and viscera |
| Thermoreceptor | Temperature change | Found in skin; separate receptors for warm and cold |
| Photoreceptor | Light | Rods and cones in retina of eye |
| Mechanoreceptor | Physical force | Touch receptors, blood pressure, stretch (in bladder and lungs) |

You may be saying to yourself, "Wait, I thought we had five senses! Have I been taught wrong my whole life?" The answer is: Not really. The five senses you're familiar with — touch, hearing, sight, smell, and taste — are your *perceived senses.* That is, you're aware that your brain has interpreted the sensory input. You say "Ouch, that's hot!" or "Wow that smells nice!" The other sensory input is, of course, processed, but that occurs outside of your conscious thought. Otherwise, you'd be constantly bombarded with thoughts about pH and blood pressure monitoring. Read on to learn about how your five perceived senses work.

# Touch

The sensory receptors all over the skin perceive at least five different types of sensation: pain, heat, cold, touch, and pressure. The five are usually grouped together as the single sense of touch.

Receptors vary in terms of their overall abundance (pain receptors are far more numerous than cold receptors) and their distribution over the body's surface (the fingertips have far more touch receptors than the skin of the back). Areas where the touch sensors are packed in are especially sensitive.

The structure of the sensory receptors varies with their function. There are *free nerve endings* (dendrites) that are responsible for itching and pain. Both the cold and warm thermoreceptors are free nerve endings as well. Other receptors are modified, where the dendrites are wrapped in connective tissue fibers forming mechanoreceptors. There are two varieties of these found in your skin. *Meissner's* (or *tactile*) *corpuscles* respond to light touch and *Pacinian* (or *lamellated*) *corpuscles* respond to a heavier, more forceful touch (pressure).

Nerve fibers that are attached to different types of skin receptors either continue to discharge during a stimulus or respond only when the stimulus starts and, sometimes, when it ends. That's why you're aware of your shoes on your feet when you first put them on, but the stimulus fades within a minute or two. In other words, *slowly adapting nerve fibers* send information about ongoing stimulation; *rapidly adapting nerve fibers* send information related to changing stimuli.

# Hearing and balance

The ear changes sound waves into nerve impulses sent to the brain (see Figure 7-6). Sound moves through air in waves of pressure. Your outer ear acts as a funnel to channel sound waves to the eardrum, causing the *tympanic membrane* (eardrum) to vibrate. (The outer ear isn't necessary for hearing, but it helps.) The ear bones, called *ossicles*, receive and amplify the vibration and transmit it to the inner ear. The vibrations create tiny ripples in the inner ear's fluid. The hollow channels of the inner ear's *cochlea* are filled with liquid, and it is lined with sensory epithelium studded with *hair cells* — mechanoreceptors that release a neurotransmitter when stimulated.

The nerve impulses travel from the left and right ears through the eighth cranial nerve to both sides of the brain stem and up to the temporal lobe — the portion of the cerebral cortex dedicated to sound.

**FIGURE 7-6:**
The anatomy
of the ear.

*Illustration by Kathryn Born, MA*

Your inner ears also transmit information to your brain regarding what position your head is in; that is, whether you're horizontal or vertical, spinning or still, moving forward or backward. Therefore, your ears are the key organ of *balance.* The process of transmitting information to the brain about your body position is basically the same as that of hearing. When you're moving, the inner ear's fluid moves and causes hair cells in the *semicircular canals* to bend, sending impulses to the brain. Your brain then processes the information about where you are spatially and initiates movements to help you keep your balance.

## Sight

Vision is probably the most complex of the senses. Your eye's *pupil,* the dot in the center of your eye that's usually black in color, allows light in. The *iris,* the pretty, colored part of your eye, contains smooth muscle that controls the pupil's size and thus the amount of light that enters the eye. The iris muscle contracts to dilate the pupil and allow more light in, such as when you're in a dark room or outside at night. The *cornea* lies anterior to the iris and pupil and merges with the *sclera,* which is the outer wall of the eye. (You can kind of see a clear area if you look at an eyeball from the side.) Both are covered by the *conjunctiva,* a mucus-like membrane. The *lens* is behind the iris and pupil (see Figure 7-7). Both the cornea and the lens help to bend light rays to aid in focusing.

**FIGURE 7-7:**
The internal
structures
of the eye.

*Illustration by Kathryn Born, MA*

A clear, gelatinous material fills the *vitreous body,* which lies behind the lens. And we're not joking when we tell you that the gelatinous material is *vitreous humor.* The transparent vitreous humor gives the eyeball its rounded shape, and it also lets light pass through it to the back of the eyeball.

The *retina* is the innermost layer of the eyeball. The retina contains two types of photoreceptors — *rods,* which detect dim light and are sensitive to motion, and *cones,* which detect color and fine details. Three types of cones detect color — one each for detecting red, blue, and green. A missing or damaged cone (regardless of the type) results in color blindness. The *macula* is the region of the retina with a high concentration of cones, providing the sharpest vision.

When light strikes the rods and cones, nerve impulses are generated and sent to cells that form the *optic nerve.* The optic nerve joins your eyeball directly to your brain and sends impulses to the brain for interpretation in the occipital lobe.

The *optic disc* is where the nerve fibers from the retina merge into the optic nerve. Thus, at this point on the retina, there are no photoreceptors — giving you a blind spot in each eye.

## Olfaction

The nose knows, but the *olfactory cells* know better. The nose is the sense organ for smell (*olfaction* is the proper term): The olfactory cells that line the top of your nasal cavity detect molecules in the air as you breathe. As you take in air through your nostrils, those molecules waft right up to your olfactory cells, where the chemicals bind to the cilia-like "hairs" that line your nasal cavity. That action initiates a nerve impulse that's sent through the olfactory cell, into the *olfactory nerve fiber*, up to the *olfactory bulb*, and right to your brain. (The olfactory bulb is the expanded area at the end of the olfactory tract where the olfactory nerve fibers enter the brain.) The brain then "knows" what the chemical odors are from, and you know what you're smelling.

## Taste

Your sense of taste *(gustation)* has a simple goal: to help you decide whether to swallow or spit out whatever is in your mouth. This extremely important decision can be made based on a few taste qualities. The tongue, the sense organ for taste, has chemoreceptors to detect certain aspects of the chemistry of food, certain minerals, and some toxins, especially poisons made by plants to deter predation by animals.

The human tongue has about 10,000 taste buds, each with between 50 to 150 chemoreceptor cells.

A taste bud's *gustatory cells* generate nerve impulses that are transmitted through the sensory nerve fiber to the gustatory areas of the brain via the seventh, ninth, and tenth cranial nerves. The brain then interprets the impulse and causes the release of the digestive enzymes needed to break down that food. So the sense of taste is tied to the endocrine system as well as to the digestive system.

Taste buds have receptors for sweet, sour, bitter, and salty sensations, as well as a fifth sensation called *umami* ("savoriness" — the flavor associated with meat, mushrooms, and many other protein-rich foods). Salty and sour detection is needed to control salt and acid balance. Bitter detection warns of foods that may contain poisons — many of the poisonous compounds that plants produce for defense are bitter. Sweet detection provides a guide to calorie-rich foods, and umami detection to protein-rich foods.

Each receptor cell in a taste bud responds best to one of the five basic tastes. A receptor can respond to the other tastes, but it responds strongest to a particular taste. And the taste buds detect only the rather unsubtle aspects of flavor. The nose detects more complex and subtler flavors.

**TECHNICAL STUFF**

The structure and function of taste buds is an area of active research and controversy, especially in the food industry. Even the exact number of basic tastes and their characterization is controversial, with new "basic tastes" being proposed from time to time. Regardless, the majority of what we perceive as flavor (good or bad) comes from olfactory input rather than gustatory.

# Pathophysiology of the Nervous System

The discussion in this section is limited to anatomical disorders of nervous tissue affecting the "lower level" functions. Major brain disorders and physiology-based psychopathologies such as schizophrenia are beyond the scope of this book.

## Chronic pain syndrome

Chronic pain is a disorder of the nervous system as much as it is a disorder of the organ causing the pain. Chronic pain has many sources (infection, trauma, and so on), but to be perceptible, the pain signal must be transmitted to the spinal cord and up to pain receptors in the brain. *Analgesics* (pain meds) work by blocking these receptors. Another approach that has proved effective in relieving chronic pain involves using a device to generate electrical impulses directly to the spinal cord to interrupt or cancel out the incoming pain impulses at the point where the spinal nerve transmitting the pain arrives at the spinal cord.

## Multiple sclerosis

*Multiple sclerosis* (MS) affects the myelin sheath that covers myelinated axons. Lesions in the myelin sheath become inflamed and irritated. When the lesions heal, scar tissue (sclerosis) on the sheath interferes with the transmission of impulses through the axon, thus blocking movement or response in the inner-vated muscle. As the disease progresses, movement becomes increasingly difficult.

## Macular degeneration

*Macular degeneration* is a vision disorder that's now a leading cause of blindness among older individuals. In macular degeneration, the *macula lutea* — a small area of the retina with a large concentration of cone photoreceptors that detect color and fine details — weakens and degenerates. Objects look smaller or larger than they really are, and colors appear faded.

One cause of macular degeneration is overgrowth of new blood vessels around the macula lutea. New growth sounds healthy, but often it's not. The new vessels leak, and as they ooze blood into the macula lutea, its delicate photoreceptors are destroyed. Macular degeneration can also result from excessive sun exposure, especially among people with blue or green eyes.

Chapter **8**

# The Endocrine System: Releasing Chemical Messages

The glands of the endocrine system and the hormones they release influence almost every cell, organ, and function of your body. The endocrine system is instrumental in regulating mood, tissue function, metabolism, and growth and development, as well as sexual function and reproductive processes.

In general, the endocrine system is in charge of body processes that happen slowly, such as cell growth. The nervous system controls faster processes like breathing and body movement. Throughout this chapter, you can see that the nervous system (which we cover in Chapter 7) and the endocrine system work closely together. The nervous system controls when the endocrine system should release or withhold hormones, and the hormones control the metabolic activities within the body. This chapter explains the functions of hormones, the glands that they come from, and common disorders of the endocrine system.

# Homing In on Hormones

A *hormone* is an *endogenous substance* (one that's produced within the body) that has its effects in specific target cells. Hormones are many and varied in their source, chemical nature, target tissues, and effects. However, they're characterized by the fact that they're synthesized in one place (gland or cell) and they travel via the blood until they reach their target cell. Hormones are bound by specific receptors on (or in) their target cells. The binding of the hormone on the receptor induces a response within the cell.

Among glands, the pituitary, thyroid, and adrenal glands are most well-known. These organs have no significant function other than to produce hormones. However, a number of other endocrine tissues and hormones, though less well-known, are just as important in controlling vital bodily functions. In fact, all the tissue in your body is, in some way, an endocrine tissue.

Hormones play a big part in *homeostasis* (see Chapter 2). When the blood passes certain checkpoints in the nervous system (such as the hypothalamus inside the brain), hormone levels are "measured." If the level of a certain hormone is too low, the gland that produces that hormone is stimulated to produce more of it. If a hormone level is too high, the gland that produces the hormone either doesn't receive any further hormonal stimulation or is "instructed" to stop or slow production. The hormone comes from the endocrine system, but the "instruction" comes from the nervous system.

REMEMBER

Your body is always keeping tabs on the metabolic processes going on inside it. If your body temperature, glucose level, or pH level leaves the normal range, the checkpoints involved in homeostasis work with the endocrine system to bring your systems back into balance.

The following sections guide you through how hormones are structured chemically, where hormones come from, and how hormones work.

## Hormone chemistry

Chemically, hormones fall into three types: those derived from lipids, from peptides, and from amines.

> » **Lipid hormones:** Lipid and phospholipid hormones are derived from fatty acids. The best-known lipid hormones are the steroids, such as estrogen, progesterone, testosterone, aldosterone, and cortisol, which are synthesized from cholesterol. Another group of lipid hormones is called *prostaglandins*.

>> **Peptide hormones:** Peptides are relatively short chains of amino acids. Peptide hormones include antidiuretic hormone (ADH), thyrotropin-releasing hormone (TRH), and oxytocin.

Other hormones are *proteins* (chains of peptides), such as insulin, growth hormone, and prolactin.

>> **Glycoprotein hormones:** More complex protein hormones bear carbohydrate side-chains and are called *glycoprotein hormones.* These include follicle-stimulating hormone (FSH), luteinizing hormone (LH), and thyroid-stimulating hormone (TSH).

>> **Amine hormones:** Amine hormones are derivatives of amino acids such as tyrosine and tryptophan. Examples are thyroxine, epinephrine, and norepinephrine.

# Hormone sources

At one time, and not so long ago, by definition a hormone was produced in an endocrine gland (and an endocrine gland was a structure that produced one or more hormones). But as biologists have discovered and described more and more hormone substances and forms, they've expanded the definition to include similar, sometimes identical, substances that have a similar mechanism of action, wherever they're produced. Check out all the sources of hormones:

>> **Endocrine glands:** An *endocrine gland* is an organ that synthesizes a hormone. It does so within a specialized cell type — the anterior pituitary gland, for instance, has cells that specialize in the production of such hormones as adrenocorticotropic hormone (ACTH), growth hormone, and TSH. Specialized cells within the thymus synthesize hormones that control the maturation of immune cells.

>> **Various organs:** A number of organs not usually included within the endocrine system by anatomists and physiologists have cells and tissues specialized for the production of hormones. For example:

- While part of the pancreas is busy secreting enzymes for the digestion of food, other specialized cells of the pancreas produce insulin, and others produce glucagon.

- The stomach and intestines synthesize and release hormones that control both physical and chemical aspects of digestion.

- Specialized cells in the ovaries and testes transform cholesterol molecules into molecules of estrogen and testosterone, respectively.

- Even the heart produces hormones, the secretion of which has an immediate strong effect on blood volume (fluid balance).

» **Neurons:** Neurons make hormones that are neurotransmitters. This seems a bit surprising, but if you think of hormones as molecules that deliver messages with considerable subtlety, it makes sense. The transmission of nerve impulses across a synapse is exactly that (flip to Chapter 7 for more on the nervous system). The only difference between epinephrine synthesized in the adrenal glands and epinephrine synthesized in nerve cells is the distance the molecules travel to their target site.

Table 8-1 lists the body's major hormones, their sources, and their functions. We provide more information about some of the individual hormones in the "Grouping the Glands" section later in this chapter.

**TABLE 8-1**     ## Important Hormones: Sources and Primary Functions

| Hormone | Source | Function(s) |
| --- | --- | --- |
| Adrenocorticotropic hormone (ACTH) | Pituitary gland (anterior part) | Stimulates secretion of corticosteroids by the cortex of the adrenal gland. |
| Antidiuretic hormone (ADH) | Pituitary gland (posterior part) | Stimulates the kidneys to reabsorb water, preventing dehydration. |
| Calcitonin | Thyroid gland | Targets the bones, kidneys, and intestines to reduce the level of calcium in the blood. |
| Epinephrine Norepinephrine | Medulla of the adrenal gland | Stimulates the heart and other muscles during the fight-or-flight response; increases the amount of glucose in the blood. |
| Estrogen | Ovaries | Stimulates the maturation and release of ova; targets muscles, bones, and skin to develop female secondary sex characteristics. |
| Glucagon | Pancreas | Causes liver, muscles, and adipose tissue to release glucose into the bloodstream. |
| Glucocorticoids | Cortex of the adrenal glands | Stimulate the formation of glucose from fats and proteins. |
| Gonadocorticoids | Cortex of the adrenal glands | Stimulate the libido. |
| Growth hormone (GH) | Pituitary gland (anterior part) | Targets the bones and soft tissues to promote cell division, and synthesis of proteins. |
| Insulin | Pancreas | Allows glucose into cells, causes liver, muscles, and adipose tissue to store glucose as a way of lowering blood glucose level. |

| Hormone | Source | Function(s) |
|---|---|---|
| Melatonin | Pineal gland | Targets a variety of tissues to mediate control of biorhythms, the body's daily routine. |
| Mineralocorticoids | Cortex of the adrenal glands | Targets the kidney cells to reabsorb sodium and excrete potassium to keep electrolytes (ions) within normal level. |
| Oxytocin | Pituitary gland (posterior part) | Social bonding, stimulates uterine contractions during childbirth and mammary glands to release milk. |
| Parathyroid hormone | Parathyroid glands | Stimulates the cells in bones, kidneys, and intestines to release calcium so that blood calcium level increases. |
| Progesterone | Ovaries | Prepares the uterus for implantation of an embryo and maintains the pregnancy. |
| Prolactin | Pituitary gland (anterior part) | Targets the mammary gland to stimulate production of milk. |
| Testosterone | Testes | Stimulates the production of sperm in testes; causes skin, muscles, and bones to develop male sex characteristics. |
| Thyroid-stimulating hormone (TSH) | Pituitary gland (anterior part) | Stimulates the thyroid gland to produce and release its important hormones, calcitonin and thyroxine. |
| Thyroxine | Thyroid gland | Distributed to all tissues to increase metabolic rate; involved in regulation of development and growth. |

# Hormone receptors

Hormones generally exit their cell of origin via *exocytosis*, which involves a sac or vesicle enveloping the substance and moving it across the cell membrane. The secreted hormone molecule goes directly into the blood and circulates until it binds with its specific receptor on the cell membrane. This activates a second messenger system, which leads to the cell's response. Or, in the case of steroid and thyroid hormones, they enter the cell to bind with the receptor inside to stimulate the response.

**REMEMBER**

The presence of a specific hormone receptor makes that cell a "target" for the hormone. Because hormones have very specific shapes, they'll only bind to their matching receptor. Without the target receptor, the hormone has no effect.

The receptor may be on or embedded in the cell membrane, as is typically the case for peptide hormones. The hormone molecule, called the *first messenger*, latches on triggering a cascade of chemical reactions within the cell. Often, the series of reactions begins with the creation of *cyclic AMP (adenosine monophosphate)*. This molecule, called the *second messenger*, then causes the target cell to produce the necessary enzymes (that is, to induce the expression of a certain gene).

A steroid hormone molecule doesn't require a cell-membrane receptor. As a lipid, it enters a cell by diffusing through the membrane; thyroid hormones enter via facilitated diffusion (see Chapter 3). After it's inside the cell, it binds with target receptor molecules either in the cytoplasm or inside the nucleus. Then, the hormone-receptor complex proceeds to its targeted gene and triggers the creation of a protein — which is the goal of the hormone.

# Grouping the Glands

In general, a *gland* is a structure that synthesizes a product that's exported from the cells. *Endocrine* (ductless) glands export their products (hormones) via the bloodstream to their target cells in anatomically distant organs. The following sections give you the lowdown on the endocrine glands. To see where these glands are located in the body, check out the "Glands of the Endocrine System" color plate in the center of the book.

**REMEMBER**

The endocrine system, like any good communications system, functions in a very integrated way. That integration complicates the discussion of which anatomical structure belongs in what category. Remember that some organs carry out multiple function which leads to their placement in multiple organ systems.

## PLEASE EXIT THROUGH THE DUCTS: THE EXOCRINE GLANDS

*Exocrine glands* (glands with ducts) produce substances that are transported through a duct to the cavity of a body organ or to the surface of the body. Most exocrine glands do not secrete hormones (the pancreas is an exception). For example, *sebaceous glands* are exocrine glands. So are sweat glands. Refer to the "Skin (Cross Section)" color plate in the center of the book for the anatomical position and structure of these glands. The sebaceous glands produce protective oils and secrete them directly onto their nearby tissues, the skin and hair. The oils don't travel around the body and don't have an effect anywhere else. The same with sweat: It's just deposited on your skin. Its only important effect, evaporative cooling, is limited to the skin immediately adjacent to the sweat gland.

Some physiologists use the term *diffuse endocrine system* to reflect the concept that many organs house clusters of cells that secrete hormones. The kidney, for example, contains scattered cells that secrete *erythropoietin*, a hormone essential for the production of red blood cells. The heart contains cells that produce *atrial naturetic hormone,* which is important in sodium and water balance.

## The taskmasters: The hypothalamus and pituitary

The *hypothalamus-pituitary complex* is the location where the nervous and endocrine systems meet. The hypothalamus and the pituitary gland adjoin in the central portion of the brain called the *diencephalon.* Under control of its "switchboard," the hypothalamus, the pituitary is the endocrine system's "master gland."

### Hypothalamus

The hypothalamus contains special cells that act as sensors that "analyze" the composition of the blood as it circulates through. It also contains other specialized cells that generate messengers (hormones) in response to the analysis. Tight pairing between these two types of cells is essential for homeostasis.

The hormones that the hypothalamus produces don't have target cells in the body. The hypothalamus synthesizes *releasing hormones* and *release-inhibiting hormones* and secretes them into small blood vessels that connect to the anterior pituitary. The pituitary responds to releasing hormones by synthesizing and releasing its hormones, which have target cells in distant organs. It responds to release-inhibiting hormones by stopping its release of the corresponding hormones. Figure 8-1 shows the relationship between the hypothalamus and pituitary glands.

### Pituitary

The pituitary gland has two parts, called the *anterior pituitary* and *posterior pituitary* that have different roles related to the hypothalamus.

The anterior pituitary gland secretes many hormones, including melanocyte-stimulating hormone (MSH). This hormone directly stimulates melanocytes to produce melanin pigment, which protects the skin from sunlight damage. This gland also secretes prolactin, which is responsible for the increase in size of the mammary glands in the breast and the production of milk.

The anterior pituitary also secretes the gonadotropic hormones FSH and LH, which target the ovaries and testicles, and ACTH, which targets the cortex of the adrenal glands. The function of these pituitary hormones is to stimulate the release of other hormones from their target glands. The same is true of growth hormone and TSH. They're messengers that stimulate the action of other endocrine glands. (This information is summarized in Table 8-1.)

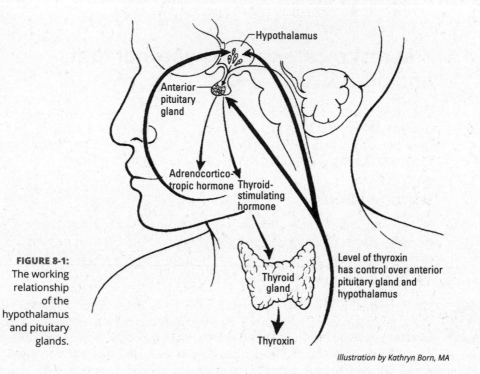

**FIGURE 8-1:**
The working relationship of the hypothalamus and pituitary glands.

Hypothalamus

Anterior pituitary gland

Adrenocortico-tropic hormone

Thyroid-stimulating hormone

Thyroid gland

Thyroxin

Level of thyroxin has control over anterior pituitary gland and hypothalamus

*Illustration by Kathryn Born, MA*

The *posterior pituitary gland* is directly connected to the hypothalamus (refer to Figure 8-1). The hormones that the posterior pituitary gland releases are actually synthesized in the nerve cell bodies of the hypothalamus. The hormones travel down the axons that end in the posterior pituitary and are released from there.

One such hormone is ADH. When the blood's fluid volume falls below the ideal range, the hypothalamus produces ADH, which travels down the axons into the posterior pituitary gland. Released by the pituitary into the blood, ADH reaches its target kidney cells. ADH binds to receptors on the cells of the tubules and alters the cells' metabolism so that more water is removed from the urine and added to the blood.

The posterior pituitary also releases *oxytocin,* the "love hormone." Its release leads to social bonding — particularly between parent and child. It also plays a major role in the progression of uterine contractions during labor, as well as *milk letdown* (release of milk from the mammary glands). In males, it has been shown to cause contraction of the spermatic duct during ejaculation.

# Controlling metabolism

Two relatively small organs exert a major effect on the availability of energy for physiological processes: the thyroid and adrenal glands.

## Thy thyroid and thou

*Thyroid hormones* affect almost every physiological process in the body. Your *thyroid gland* looks somewhat like a butterfly that straddles your *trachea* (windpipe). (It's pictured in Figure 8-1 and in the "Glands of the Endocrine System" color plate in the center of the book.) Each lobe of the thyroid gland — the butterfly's wings — is adjacent to the trachea; a stretch of tissue called the *isthmus* connects the lobes. Simple cuboidal epithelium lines the follicles within the lobes and secretes a jellylike substance called *thyroglobulin*. Thyroglobulin "traps" iodine ions (consumed in food) in the *colloid* (internal fluid of the follicle) and promotes the formation of the amine hormones *thyroxine (T4)* and *triiodothyronine (T3)*. When TSH from the anterior pituitary binds the target receptors in the thyroid, the thyroid hormones are released slowly into the bloodstream.

The thyroid hormones regulate the following physiological reactions. We discuss some of the problems that arise from the malfunction of these hormones in the "Pathophysiology of the Endocrine System" section later in the chapter.

The thyroid hormones

>> Control the body's *basal metabolic rate* (the amount of energy needed to keep the body function at rest).

>> Increase the rate at which cells use glucose for energy.

>> Help maintain body temperature by increasing or decreasing metabolic rate.

>> Regulate growth and differentiation of tissues in children and teens.

>> Increase the amount of certain enzymes in the mitochondria that are involved in oxidative reactions.

>> Influence the breakdown rate of proteins, fats, carbohydrates, vitamins, minerals, and water.

>> Stimulate mental processes.

>> Increase the rate of protein synthesis.

## Topping the kidneys

The root *renal* means "kidney"; the prefix *ad* means "near." So the adrenal glands are near the kidneys. More specifically, they sit on top of the kidneys (see Figure 8-2 and refer to the "Glands of the Endocrine System" color plate in the center of the book). Like the skin on a kidney bean, a thin *capsule* covers the entire adrenal gland. Inside, each adrenal gland has two parts: the *cortex* (outer layer) and the *medulla* (middle portion), which have different functions.

**FIGURE 8-2:**
The location of the adrenal glands.

*Illustration by Kathryn Born, MA*

## ADRENAL CORTEX

The adrenal cortex secretes *corticosteroids*, which include *mineralocorticoids, glucocorticoids,* and *gonadocorticoids* (refer to Table 8-1). One of the most important mineralocorticoids is *aldosterone*, which is responsible for regulating the concentration of *electrolytes*, such as potassium ($K^+$), sodium ($Na^+$), and chloride ($Cl^-$) ions. This regulation keeps the blood's salt and mineral content within the ranges required for homeostasis.

*Electrolytes* are substances that split apart into *ions* (atoms or molecules with a positive or negative charge) when in solutions, such as the watery tissue fluid around your cells or the cytoplasm within your cells. As their name suggests, electrolytes are capable of conducting electricity.

Aldosterone targets the kidneys' tubules and stimulates the resorption of sodium ions. When the sodium ions are reabsorbed into the bloodstream, chloride ions quickly follow. $Na^+$ and $Cl^-$ love to be together as NaCl, commonly known as "salt." And where salt goes, water follows. If salt ions move into the bloodstream, water does too, increasing the blood's fluid volume. Ultimately, the fluid and electrolyte balance affects blood pressure (see Chapter 9 for everything you wanted to know about the cardiovascular system).

The *gonadocorticoids* were named such because they are identical to the steroid hormones made by the gonads, the testicles, and ovaries. The gonadocorticoids consist of *testosterone, estrogen,* and *progesterone.* If you thought that as a woman you didn't have any testosterone or as a man you didn't have any estrogen or progesterone, you're wrong. Admittedly, these are secreted in small amounts and have little influence on the development of the reproductive system. Apparently, their primary responsibility is to heighten sex drive. Increased production of their usually low amounts can lead to *feminization* in the male and *masculinization* in the female.

*Cortisol,* the main glucocorticoid hormone, regulates the metabolism of proteins, fats, and carbohydrates. Your body releases cortisol when you're stressed emotionally, physically, or environmentally (hence its being dubbed the "stress hormone"). Cortisol affects metabolism in the following ways:

>> It breaks down protein, decreases protein synthesis, and moves amino acids from tissues to liver cells to promote *gluconeogenesis* (creation of glucose) and the formation of glycogen.

>> It moves fat from adipose tissue to the blood.

>> It reduces the rate at which cells take in glucose.

Cortisol and other corticosteroids affect the immune system by decreasing the number of circulating immune cells and decreasing the size of the lymphoid tissue. It also acts as an anti-inflammatory. Under severe stress and with large amounts of glucocorticoids circulating in the blood, the lymphoid tissue is unable to produce antibodies (see Chapter 13 for more on the lymphatic system). The role of corticosteroids in susceptibility to infectious disease is an active area of medical research.

## ADRENAL MEDULLA

The adrenal medulla developed from the same tissues as the sympathetic nervous system (we cover the nervous system in Chapter 7). Some of the adrenal medulla's functions involve regulating actions of structures of the sympathetic nervous system, including a class of hormones called the *catecholamines,* of which *epinephrine* and *norepinephrine* are the best known.

Epinephrine, also called *adrenaline,* initiates the "adrenaline rush" of the *fight-or-flight response.* It stimulates the release of free fatty acid molecules from your adipose tissue. Your muscles — including your heart muscle and respiratory system muscles — use these fatty acid molecules for energy, saving glucose for use by your brain. After all, if you're fighting or fleeing, you need to think.

Unlike most hormones, epinephrine produces its effect almost instantaneously.

Like epinephrine, norepinephrine is a catecholamine that's tied to the nervous system. Norepinephrine causes *vasoconstriction* — that is, tightening of the blood vessels. Norepinephrine is released to increase blood pressure when the hypothalamus senses *hypotension* (low blood pressure), as well as when you're stressed. (Your body still prepares to run or fight, even when these responses aren't quite appropriate.)

## Getting the gonads going

Your *gonads* — *ovaries* if you're female or *testes* if you're male — produce and secrete the steroid sex hormones — *estrogen* and *progesterone* in females and *testosterone* in males. Your body secretes sex hormones throughout your lifetime at different levels. Their production increases at puberty and normally decreases as you age.

You may think that estrogen is limited to female animals, but you'd be wrong. Estrogen can be detected in the urine of male animals, and even, surprisingly, in growing plants! Just as men also have some estrogen, women also have some testosterone.

### Estrogen

In women, the increased production of estrogen at puberty is responsible for initiating the development of the secondary sex characteristics, such as the enlargement of the breasts. Bone tissue grows rapidly, and height increases. Estrogen helps this process, causing calcium and phosphate to be transported in the bloodstream so they can be used for bone growth and for stimulating the activity of the osteoblasts (see Chapter 5).

Estrogen also allows the pelvic bones to widen to allow passage of an infant during childbirth. In addition, estrogen increases the deposition of fat around the body, thus giving women a more rounded appearance than men.

TECHNICAL
STUFF

Estrogen is actually a category — referring to the steroid hormones made by the ovaries. Use of the term refers to all three variants, the most common of which is *estradiol.*

# Skin (Cross Section)

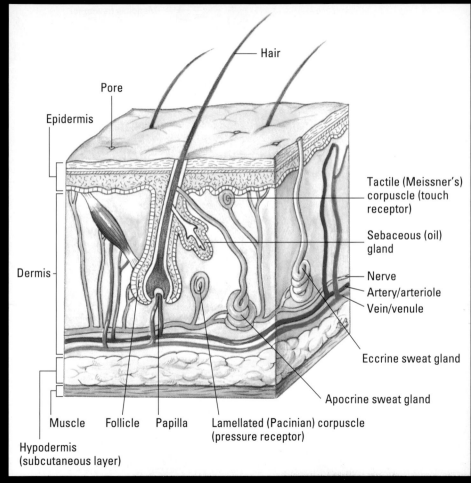

The skin's many layers protect the body from the environment.
See Chapter 4.

# Major Bones of the Skeleton

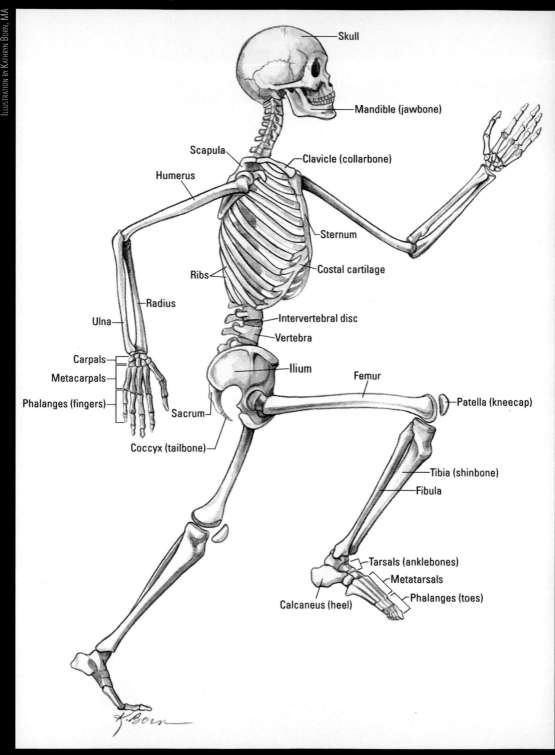

Skull

Mandible (jawbone)

Scapula

Clavicle (collarbone)

Humerus

Sternum

Costal cartilage

Ribs

Radius

Intervertebral disc

Ulna

Vertebra

Carpals

Metacarpals

Ilium

Femur

Phalanges (fingers)

Patella (kneecap)

Sacrum

Coccyx (tailbone)

Tibia (shinbone)

Fibula

Tarsals (anklebones)

Metatarsals

Phalanges (toes)

Calcaneus (heel)

The skeleton comprises the bones and the joints that connect them. See Chapter 5.

# Muscular System

ILLUSTRATION BY KATHRYN BORN, MA

Temporalis

Trapezius

Sternocleidomastoid

Deltoid

Pectoralis major

Triceps

Biceps

Brachialis

Flexor muscles of the forearm

Serratus anterior

Latissimus dorsi

External oblique

Extensor muscles of the forearm

Rectus abdominis

Quadriceps

Illotibial tract

Gluteus maximus

Gracilis

Semitendinosus

Biceps femoris

Gastrocnemius

Fibularis longus

Semimembranosus

Tibialis anterior

Sartorius

Soleus

K.Born

**The muscular system works with the skeletal and nervous systems to move the body both spatially and internally. See Chapter 6.**

# Nervous System

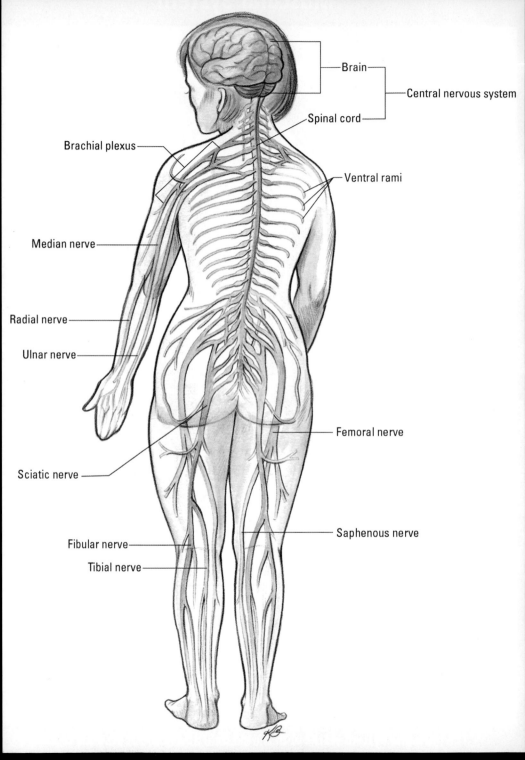

ILLUSTRATION BY KATHRYN BORN, MA

Brain

Central nervous system

Spinal cord

Brachial plexus

Ventral rami

Median nerve

Radial nerve

Ulnar nerve

Femoral nerve

Sciatic nerve

Saphenous nerve

Fibular nerve

Tibial nerve

**The nervous system comprises the central nervous system and the peripheral nervous system. See Chapter 7.**

# Glands of the Endocrine System

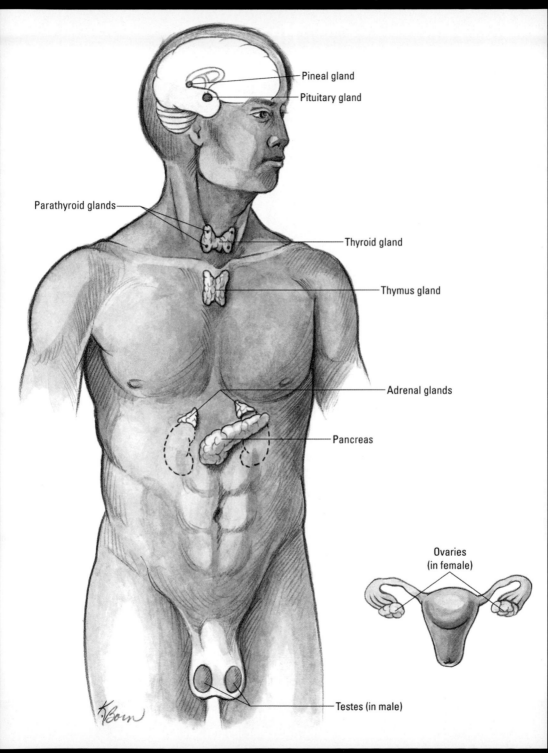

Pineal gland

Pituitary gland

Parathyroid glands

Thyroid gland

Thymus gland

Adrenal glands

Pancreas

Ovaries
(in female)

Testes (in male)

**Endocrine glands produce hormones that are distributed throughout the body in the blood. See Chapter 8.**

# Heart

Aorta

Pulmonary artery

Superior vena cava

Left atrium

Aortic valve

Left atrioventricular
(bicuspid or mitral) valve

Right atrium

Left ventricle

Right atrioventricular
(tricuspid) valve

Chordae
tendineae

Apex of the heart

Inferior vena cava

Right ventricle

Papillary muscle

ILLUSTRATION BY KATHRYN BORN, MA

K. Born

**Effective blood circulation depends on the functioning of the heart's interior
structure and its outer muscular layers. See Chapter 9.**

# Arterial Components of the Cardiovascular System

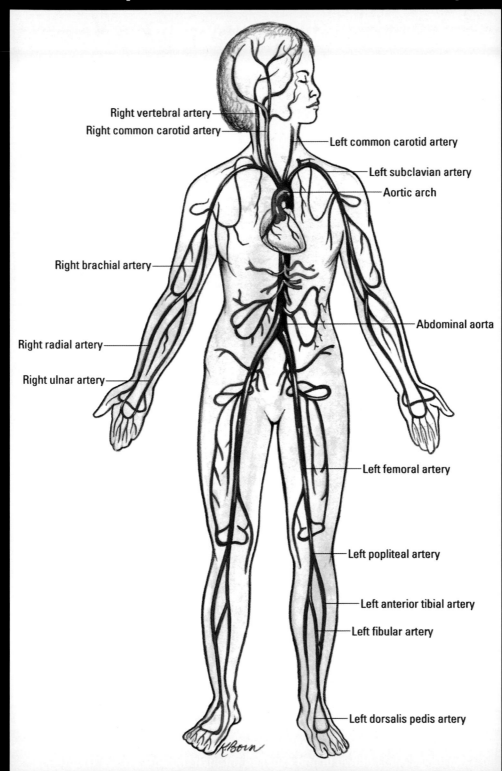

Right vertebral artery

Right common carotid artery

Left common carotid artery

Left subclavian artery

Aortic arch

Right brachial artery

Abdominal aorta

Right radial artery

Right ulnar artery

Left femoral artery

Left popliteal artery

Left anterior tibial artery

Left fibular artery

Left dorsalis pedis artery

The arteries carry oxygenated blood from the heart to all parts of the body.
Deoxygenated blood returns to the heart via the veins (not shown). See Chapter 9.

# Respiratory System

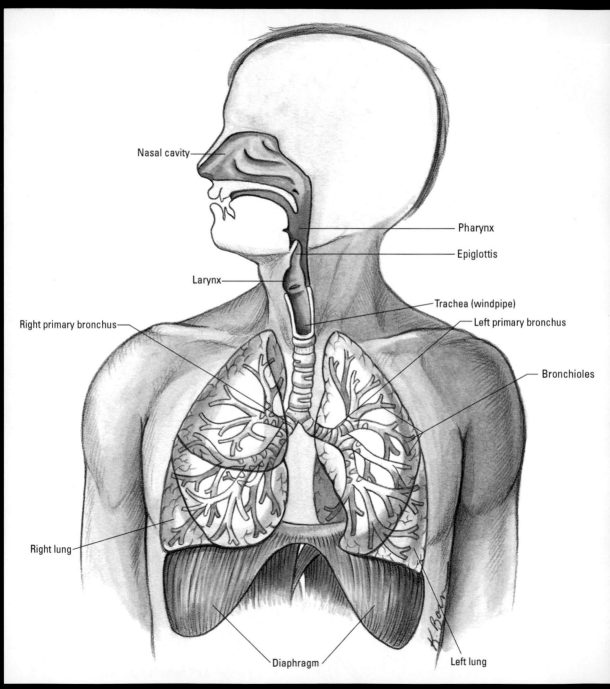

Nasal cavity

Pharynx

Epiglottis

Larynx

Trachea (windpipe)

Right primary bronchus

Left primary bronchus

Bronchioles

Right lung

Diaphragm

Left lung

**Contraction and release of the diaphragm alternately decreases and increases air pressure in the lungs. Air is drawn in and expelled through the airway. See Chapter 10.**

ILLUSTRATION BY KATHRYN BORN, MA

Terminal bronchiole

Pulmonary venule

Pulmonary arteriole

Lymphatic vessel

Respiratory bronchiole

Smooth muscle

Pulmonary capillary

Alveolar duct

Alveoli

Alveolar sac

The pulmonary capillary is the specific site of gas exchange. See Chapter 10.

# Digestive System

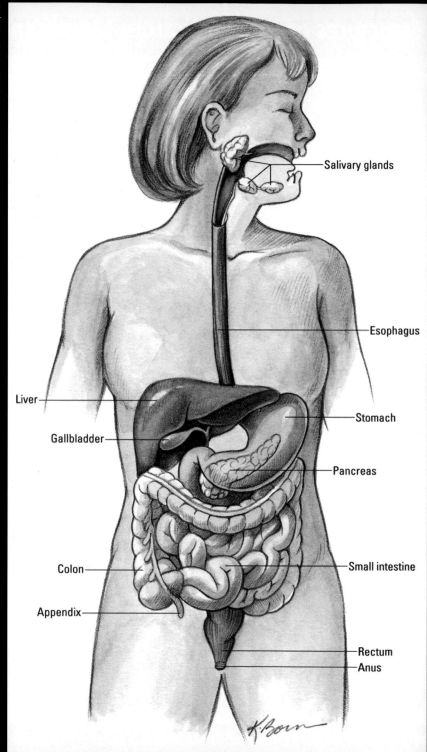

Salivary glands

Esophagus

Liver

Stomach

Gallbladder

Pancreas

Colon

Small intestine

Appendix

Rectum

Anus

**The transformation of food into physiologically available nutrients involves the participation of many organs. See Chapter 11.**

# Stomach

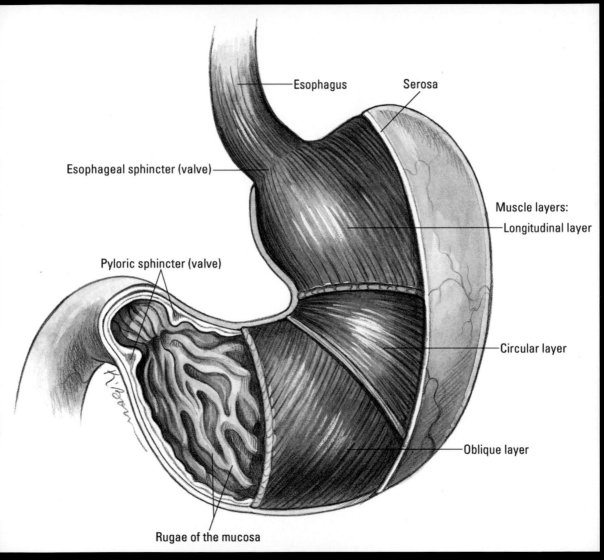

Esophagus

Serosa

Esophageal sphincter (valve)

Muscle layers:

Longitudinal layer

Pyloric sphincter (valve)

Circular layer

Oblique layer

Rugae of the mucosa

ILLUSTRATION BY KATHRYN BORN, MA

**The stomach's tissue layers are a variation on a pattern that repeats all along the digestive tract: connective tissue, smooth muscle layers, and mucosa. See Chapter 11.**

# Urinary System

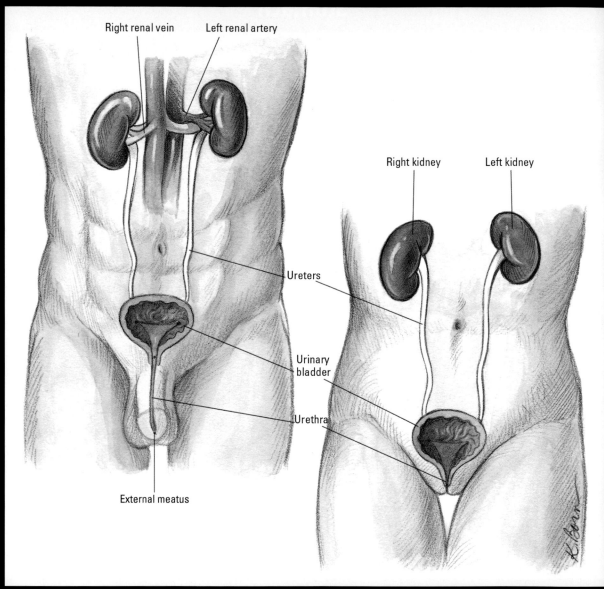

Right renal vein    Left renal artery

Right kidney    Left kidney

Ureters

Urinary bladder

Urethra

External meatus

**The urinary system is specialized for the elimination of wastes and toxins. Its most complex organ, the kidney, also performs many other high-level homeostatic functions. See Chapter 12.**

# Kidney and Nephron

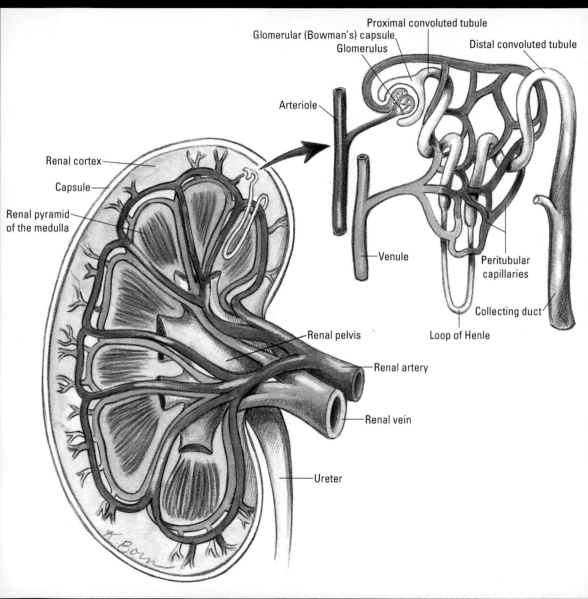

Illustration by Kathryn Born, MA

Proximal convoluted tubule

Glomerular (Bowman's) capsule

Glomerulus

Distal convoluted tubule

Arteriole

Renal cortex

Capsule

Renal pyramid of the medulla

Venule

Peritubular capillaries

Renal pelvis

Collecting duct

Loop of Henle

Renal artery

Renal vein

Ureter

**The kidney is specialized for chemical processing, distribution, and disposal. The nephron is the kidney's filtering unit. See Chapter 12.**

# Lymphatic System

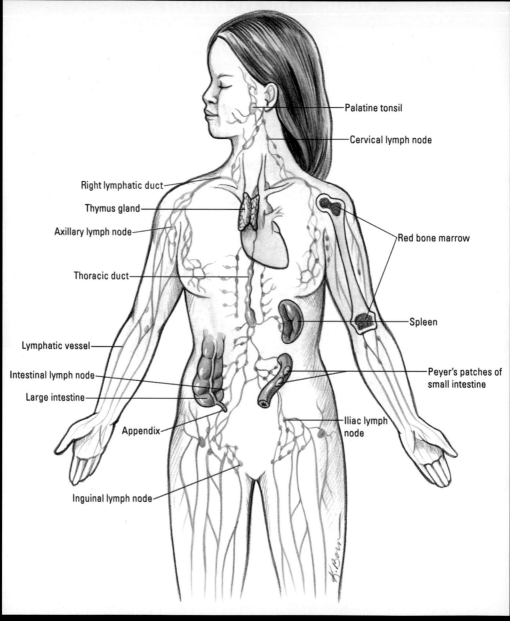

Palatine tonsil

Cervical lymph node

Right lymphatic duct

Thymus gland

Axillary lymph node

Thoracic duct

Red bone marrow

Lymphatic vessel

Intestinal lymph node

Large intestine

Spleen

Peyer's patches of small intestine

Appendix

Iliac lymph node

Inguinal lymph node

**The lymphatic system forms the infrastructure of the body's immune surveillance system. See Chapter 13.**

# Reproductive System (Female and Male)

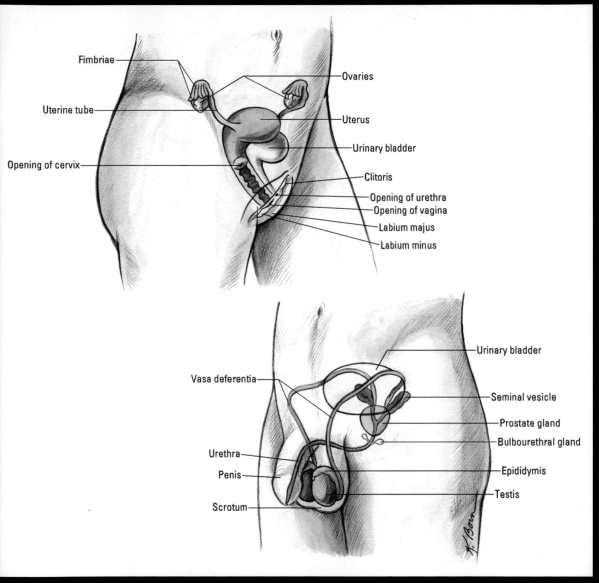

Fimbriae
Ovaries
Uterine tube
Uterus
Urinary bladder
Opening of cervix
Clitoris
Opening of urethra
Opening of vagina
Labium majus
Labium minus

Urinary bladder
Vasa deferentia
Seminal vesicle
Prostate gland
Bulbourethral gland
Urethra
Penis
Epididymis
Testis
Scrotum

Female and male reproductive anatomy are complementary to fulfilling the
evolutionary imperative of reproduction with variation. See Chapter 14.

# Prenatal Development

ILLUSTRATION BY KATHRYN BORN, MA

**Embryo at 5 Weeks**
Length: Less than ½ inch

Head

Optic cup

Umbilical cord

Upper limb bud

Vertebral column

Lower limb bud

Tail

**Embryo at 7 Weeks**
Length: About 1 inch

Ear bud

Olfactory pit

**Fetus at 9 Weeks**
Length: About 3 inches

**Fetus at 12 Weeks**
Length: About 3½ inches

**Fetus at 21–25 Weeks**
Length: 11–15 inches

**Infant at Birth**
35–38 weeks
Length: 21 inches

**Rapid growth and differentiation in the fetus are fueled through the placenta.
See Chapter 15.**

## Progesterone

Progesterone works to prepare the uterus for implantation of a pre-embryo by causing changes in uterine secretions and in storing nutrients in the uterus's lining. Progesterone also contributes to breast development.

## Testosterone

Testosterone causes the development of secondary sex characteristics in males. As a boy hits puberty, his muscle tissues start to grow, his sex organs enlarge, hair develops on his chest and face, and the hair on his arms and legs becomes darker and coarser. For more on development during puberty, turn to Chapter 15.

These hormones also play a key role in the development of *gametes*, or sex cells (egg and sperm). This is discussed in Chapter 14.

# THE SHORT BUT ONGOING HISTORY OF HORMONE REPLACEMENT THERAPY

As they age, men and women produce fewer sex hormones. In a man, the decline typically begins in the 30s and proceeds slowly. Many men retain their sexual and reproductive functions late in life. For women, the decline begins in the 30s, as with men, but becomes precipitous in the late 40s or 50s. This is the period of *menopause.*

The effects of sex hormone depletion are similar in both sexes, but they happen more suddenly in women. An early effect in women is the cessation of reproductive function — ovulation ceases and pregnancy is no longer possible. Faulty signaling in the parasympathetic nervous system causes irritability and disrupts the body's temperature-monitoring ability, causing hot flashes. Chronically, though, cellular metabolism slows down. *Bone remodeling* (the constant buildup and breakdown of bone tissue) tilts in the direction of breakdown. Replacement of the structural proteins in the skin slows to a crawl.

Because all of these undesirable effects could be shown to be related to decreased hormone levels, researchers questioned whether they could develop hormone therapy to slow down or stop some of this creeping decrepitude. Between the 1960s and the 1990s, various types of female hormone-replacement drugs became available at low costs. In the 1980s and 1990s, primary care physicians offered estrogen-replacement therapy routinely to women in menopause. Some 20 percent of women in menopause took hormone replacement therapy (HRT), a huge patient population. Many women, it seemed, hoped to enjoy their post-reproductive years in health and vitality.

*(continued)*

*(continued)*

Their expectations were based on some early research that had indicated beneficial effects on the cardiovascular and nervous systems of women who chose HRT. As is routine in research, larger studies were designed to test the results of these early studies. Exemplary data for this study even existed, from the Women's Health Initiative (WHI), a large, randomized, clinical trial, sponsored by the National Institutes of Health that included more than 16,000 healthy women.

The medical "outcomes" in women who underwent HRT surprised nearly everyone when they were published in 2002. Although researchers expected some additional cancers from increased estrogen, the numbers were disturbing. Researchers had expected to see a "cardioprotective effect," based on the observation that heart disease remains low in women until menopause. But cardiovascular diseases were actually more prevalent and more severe in those who had taken HRT. The therapy had minimal effect on bone retention (exercise was better), which ceased immediately on discontinuation of the therapy. It didn't have a measurable effect on cognition or memory. Its only reliable effect, it seemed, was the diminution of hot flashes in the first months or years after menopause (not dismissed lightly by anyone who has suffered them). Follow-up analyses of these same data have supported the original conclusions.

Women's attitudes and doctors' prescribing habits changed abruptly. Though the current prescribing guidelines are considered safe, as supported by the data in the WHI study, some women are choosing to endure hot flashes.

Interest in the clinical uses of hormones to control the symptoms of aging remains strong, but the experience so far suggests that the endocrine function is more subtle and entwined than researchers imagined.

# Enteric endocrine

Much of the endocrine function is *enteric* (related to digestive processes). Precise control of nutrient intake and storage and the excretion of toxins and digestive byproducts is essential for homeostasis and metabolism.

## Flat as a pancreas

Your *pancreas* is a fibrous, elongated, flat (as a pancake) organ that lies nestled in your abdomen near your kidneys, stomach, and small intestine (see Figure 8–3 and the "Glands of the Endocrine System" color plate in the center of the book). Its two tissue types have different functions:

>> **Digestive function:** It produces digestive enzymes that it secretes into the small intestine. This tissue is arranged around ducts (see Chapter 11).

>> **Endocrine function:** It produces the hormones *insulin* and *glucagon,* which it secretes into the blood. This tissue is arranged in clusters called islets (see Figure 8-3).

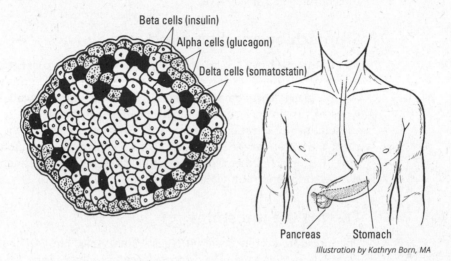

Beta cells (insulin)
Alpha cells (glucagon)
Delta cells (somatostatin)

Pancreas    Stomach

*Illustration by Kathryn Born, MA*

**FIGURE 8-3:** Anatomy of the pancreas.

Glucose travels in the blood, making it available as an energy molecule for all cells. But too much glucose in the blood is harmful to the small vessels, especially those in the extremities, in the kidneys, and in the retina of the eye. The body must keep the concentration of glucose in the blood within limits. The pancreas's endocrine tissues produce insulin and glucagon, which work together to balance the blood's glucose level — insulin acts to lower it, and glucagon acts to raise it. The regulation of blood glucose is a classic example of a hormone mechanism of homeostasis. Refer to the "Pathophysiology of the Endocrine System" section later in this chapter for a discussion of disorders in this system.

**TECHNICAL STUFF**

Insulin secretion is an example of a *negative feedback mechanism.* As the blood's glucose level declines, the body slows the secretion of insulin until the next surge of blood glucose following a meal or snack.

Insulin is released when the blood's glucose level rises. It acts by stimulating glucose uptake in cells. Without insulin, most cells would not be able to take in glucose, which is required for cellular respiration (see Chapter 2). So, in addition to the damage done by excess glucose in the blood, cells cannot generate enough ATP to power their processes. Insulin also stimulates glucose uptake activity in the energy storage cells in the liver, muscle cells, and adipose (fatty) tissue.

A low blood glucose level stimulates the pancreas to secrete insulin's partner *glu-cagon*, which pulls glucose from cells where it's stored and releases it into the blood. In particular, it stimulates the liver to break down *glycogen*, the storage form of glucose, leading to more sugar in the blood. Glucagon keeps the metabolic fires burning at a steady level.

## Stomaching another gland

Yes, the stomach is a key organ in digestion (Chapter 11 covers the digestive system in detail), but it also secretes hormones that are used during digestion, making it a gland. The stomach secretes a group of hormones called *gastrins*. Many types of gastrin molecules — small, medium, and large — are responsible for stimulating the secretion of gastric acid. Gastrins also control the sphincter muscle at the bottom of the esophagus, thereby controlling when food can be passed into the stomach. Other cells in the stomach produce the hormone *ghrelin*, which targets the brain to stimulate the hunger.

## Testing the intestines

The intestines secrete powerful digestive enzymes, but they require a pH much higher (less acidic) than the extremely acidic stomach acids. The small intestine produces the hormone *secretin* that stimulates the release of neutralizing substances, such as bile from the gallbladder and pancreatic bicarbonate. Other cells in the small intestine produce *cholecystokinin* (CCK), which triggers the release of digestive enzymes from the pancreas and gall bladder.

# Other endocrine glands

Other endocrine glands secreting important hormones include the following:

>> **Parathyroid:** The *parathyroid glands* are four small glands that secrete *parathyroid hormone* (PTH), which increases the concentration of calcium in the blood, making it available to the muscle fibers and neurons. The parathyroid essentially helps the nervous and muscular systems function properly. Calcium is the primary element that causes muscles to contract, and calcium levels are very important to the communication between neurons.

The four parathyroids are typically found on the backside of the thyroid. They're about the size and shape of a grain of rice. They are related to the thyroid in their function of maintaining calcium levels, as the thyroid produces the hormone *calcitonin,* which decreases blood calcium levels.

>> **Pineal gland:** The *pineal gland* is a small oval gland in the brain between the cerebral hemispheres that's considered part of the epithalamus of the diencephalon. The pineal gland secretes the hormone *melatonin,* which plays a role in regulating the body's *circadian rhythm* — the normal fluctuation of physiology over the day-night cycle. The secretion of melatonin is influenced by the perception of light within the gland.

>> **Thymus:** The *thymus* is a lobed gland situated in the thoracic cavity, just below the collarbones and just above the heart. The thymus's main function is to stimulate the maturation of T lymphocytes from the bone marrow into T cells (see Chapter 13). The thymus produces a group of hormones called *thymosins* that are involved in differentiating and stimulating the immune system's cells.

# Pathophysiology of the Endocrine System

The body depends on its chemical messaging system to maintain control over its physiological processes. Malfunctions in the messaging system can disrupt target organ systems and the cardiovascular system that transports the hormones between the gland and the target organs via the blood.

## Abnormalities in insulin metabolism

The following abnormalities in insulin metabolism cause blood glucose to rise and remain at high levels. This extra glucose damages the smallest blood vessels, like those in the retina of the eye and the glomeruli of the kidney. It can also lead to acidosis, a drop in the pH of the blood, which damages numerous tissues including nerves.

### Metabolic syndrome

Insulin moves glucose into cells. However, sometimes the cells develop a "resistance" to insulin's effects, and larger and larger concentrations of insulin are required to produce the same results. As long as the pancreas is producing enough insulin to overcome this resistance, blood glucose levels remain within the homeostatic range. Eventually, the pancreas can no longer produce enough insulin to overcome the resistance, initially after meals, when blood glucose levels are highest. Insulin resistance is thought to contribute to the accumulation of abdominal fat and other developments that increase risk for diabetes and cardiovascular disease. This set of risk factors is called the *metabolic syndrome.*

## Diabetes mellitus type 1

*Type 1 diabetes*, previously called *juvenile diabetes*, is the result of the destruction of the pancreatic cells that produce insulin. The ultimate cause of the cell destruction has been an active area of medical research for a century. The effects of untreated type 1 are ultimately fatal. The development of intravenous insulin for treatment of this condition was one of the medical breakthroughs of the last century.

## Diabetes mellitus type 2

*Type 2 diabetes* is caused by insufficient blood insulin concentration, either because insulin production is impaired or insulin resistance has developed. The high levels of blood glucose slowly damage small blood vessels, leading to impairment or failure of numerous organs and systems.

## Diabetes insipidus

*Diabetes insipidus* is caused by the inability of the hypothalamus to produce the proper amount of antidiuretic hormone (ADH), which is responsible for stimulating the kidney to return water to the bloodstream. Without ADH, very little water is returned, and the concentration of glucose in the blood rises (along with that of other dissolved substances). Large amounts of watery urine lead to dehydration and thirst and carry electrolytes right out of the body. This disorder can be treated by administration of ADH therapy.

## Gestational diabetes

*Gestational diabetes* is high blood glucose that develops at any time during pregnancy in a woman who doesn't have diabetes. Women who have gestational diabetes are at high risk for type 2 diabetes and cardiovascular disease later in life.

# Thyroid disorders

The pervasive role of the thyroid hormones in metabolism can be seen by the widespread physiological effects of thyroid disorders.

## Hypothyroid disorders

The prefix *hypo* means "below." The result of low levels of the thyroid hormones is *hypothyroidism*. This low level can result from a defect in the thyroid gland (*primary hypothyroidism*), or the hypothalamus or pituitary glands may not be sending the proper messaging hormones to the thyroid (*secondary hypothyroidism*). People with primary hypothyroidism may have inflammatory conditions similar to arthritis, or chronic conditions, such as *Hashimoto's thyroiditis* (a disease in which the body's immune system attacks the thyroid gland's cells). Dietary deficiency of iodine and medications that negatively affect the thyroid gland also may cause secondary hypothyroidism.

Hypothyroidism has many signs and symptoms because the thyroid hormones have such a widespread effect. Nearly every cell in the body is stimulated by the hormone *thyroxine*, which regulates the rate of metabolism. Symptoms of hypothyroidism are shown in Table 8-2.

**TABLE 8-2**     ## Symptoms of Hypothyroidism

| Initially | As Disease Progresses | Severe |
| --- | --- | --- |
| Fatigue | Decreased sex drive | Psychiatric problems; changes in behavior |
| Cold sensitivity | Stiff joints | Carpal tunnel syndrome |
| Weight gain without increase in food intake or decrease in exercise | Muscle cramps | High cholesterol, poor circulation, heart problems |
| Constipation | Numbness or tingling sensations | Dry skin and hair, some hair loss; brittle, grooved nails |
| Memory problems | | Impaired fertility |
| | | Weak colon, intestinal obstruction, anemia |

*Myxedema* is a life-threatening complication of hypothyroidism. As metabolism slows, the exchange of carbon dioxide and oxygen slows. As the amount of carbon dioxide in the blood rises, the patient is at risk of slipping into a coma, which may be fatal.

Whether primary or secondary, hypothyroidism has profound effects on metabolism in several organ systems (refer to Table 8-2). Treatment for people with hypothyroidism involves lifelong administration of a synthetic thyroid hormone. However, therapy must begin gradually so the heart isn't negatively affected.

## Hyperthyroid disorders

Abnormally high levels of the thyroid hormones is the condition of *hyperthyroidism*, also called *Graves' disease*. Hyperthyroidism makes a person irritable, nervous, and unable to sleep. The thyroid gland may enlarge into a goiter, and swelling of the eye muscles may make the eyeballs protrude somewhat. Treatment options for patients with hyperthyroidism include oral medications, a single dose of radioactive iodine, or surgery to reduce the size and activity of the thyroid gland.

# Androgen insensitivity

*Androgen insensitivity syndrome* (AIS) is a disorder caused by mutation of the gene for the receptor that binds testosterone, which regulates the expression of genes that stimulate male sexual development. Affected individuals are chromosomally XY but have a feminine phenotype and are sterile. AIS completely or partially prevents development of male sexual characteristics in the fetus, despite the presence of the Y chromosome. The extent of the syndrome ranges from complete androgen insensitivity and development of normal external (but not internal) female sexual anatomy, to partial insensitivity with altered or ambiguous male or female genitals, to mild insensitivity with normal male genitals, enlarged breasts, and possibly impotence.

# 4

# Exploring the Inner Workings of the Body

IN THIS CHAPTER

» Looking at your blood and what's in it

» Discovering arteries, veins, and capillaries

» Breaking down the heart's parts

» Following blood along its path through your body

» Looking at some cardiovascular system problems

Chapter **9**

# The Cardiovascular System: Getting Your Blood Pumping

More than any other system, the cardiovascular system has contributed strong imagery to people's daily language. "Heart" is a metaphor for love and for courage. People say they "nearly had a heart attack" to describe an experience of surprise or shock. Abstract but distinctive characteristic qualities are said to be "in one's blood." The blood itself runs cold, runs hot, and runs all over the place in informal speech and poetry. Scientifically speaking, emotions are more a matter of hormones than myocardium, and nobody's blood is any redder or any hotter than anyone else's. The heart is neither soft nor hard, but a muscular, fibrous pump, and blood is a complex biological fluid that must be kept moving through its specialized network of vessels. Ready to take a closer look?

# Getting Substances from Here to There

The functions of the cardiovascular system are all related to transportation. Nearly every substance made or used in the body is transported in the blood: hormones, gases of respiration, products of digestion, metabolic wastes, and immune cells. We discuss these transportation functions in the context of the other organ systems only as they relate specifically to cardiovascular anatomy and physiology.

The blood also "transports" heat. Stimulated by the hormones of thermoregulation, blood flow can disperse heat to the environment at the body surface or conserve heat for essential functions in the body core.

# Carrying Cargo: Your Blood and What's in It

Blood — that deep red, body-temperature-warm, and metallic-tasting liquid that courses through your body — is a vitally important, life-supporting, life-giving, life-saving substance that everybody needs. And every adult-size body contains about five liters (1.3 gallons) of the precious stuff.

Blood consists of many different types of cells in a matrix called *plasma*. That's what makes blood a connective tissue. The different types of cells — *red blood cells*, *white blood cells*, and *platelets* — are referred to as *formed elements*. They combine to 45 percent of the blood volume; the remaining 55 percent is plasma.

## Watering down your blood: Plasma

Plasma is about 92 percent water. The remaining 8 percent or so is made up of *plasma proteins*, salt ions, oxygen and carbon dioxide gases, nutrients (glucose, fats, amino acids) from the foods you take in, *urea* (a waste product), and other substances carried in the bloodstream, such as hormones and enzymes.

The plasma proteins, produced in the liver, are made for the plasma. That is, they aren't being transported anywhere. Their functions include

>> **Albumin:** The smallest plasma protein and the most abundant, albumin maintains the *osmotic pressure* (tendency to pull water in) in the bloodstream within the homeostatic range.

>> **Fibrinogen:** During the process of clot formation, fibrinogen is converted into threads of *fibrin,* which then form a meshlike structure that traps blood cells to form a clot.

>> **Immunoglobulin:** This is another word for *antibody* — proteins that are created in response to an invading microbe (turn to Chapter 13 for more on the immunity).

## Transporting oxygen and carbon dioxide: Red blood cells

Red blood cells (RBCs), or *erythrocytes* (*erythro* is the Greek word for "red"), are the most numerous of the blood cells and one of the most numerous of all cell types in your body. About one-quarter of the body's approximately 3 trillion cells are RBCs. RBCs are among the cell types that must be constantly regenerated and disposed of. In fact, you produce and destroy a few million RBCs every second!

The cytoplasm of RBCs is full to the brim with an iron-containing biomolecule called *hemoglobin.* The iron-containing heme group in hemoglobin binds oxygen at the respiratory membrane (see Chapter 10) and then releases it in the capillaries. This is the sole mechanism by which all your cells and tissues get the oxygen they need to sustain their metabolism. RBCs containing heme-bound oxygen are bright red, the familiar color of the arterial blood that flows from wounds. RBCs in the venous system have less heme-bound oxygen and are a dark red.

**TECHNICAL STUFF**

RBCs are so full of hemoglobin because they contain very little else. During differentiation, they lose their organelles, even their nucleus. Only their membranes retain their function.

An RBC has about a four-month life span, at the end of which it's destroyed by a *phagocyte* (a large cell with cleanup responsibilities) in the liver or spleen. The iron is removed from the heme group and is transferred to either the liver (for storage) or the bone marrow (for use in the production of new hemoglobin). The rest of the heme group is converted to *bilirubin* and released into the plasma (giving plasma its characteristic straw color). The liver uses the bilirubin to form bile to help with the digestion of fats.

At the same time as oxygen diffuses into a cell, carbon dioxide diffuses out, making its way in the interstitial fluid to the venous system. Some deoxygenated hemoglobin in the venous blood takes up carbon dioxide to form *carboxyhemoglobin.* At the respiratory membrane, carboxyhemoglobin releases the carbon dioxide and takes up oxygen again. Carbon dioxide is transported in several different ways in the blood to the respiratory membrane, where it enters the lung and is exhaled in the breath.

## Plugging along with platelets

*Platelets* are tiny pieces of cells. Large cells in the red bone marrow called *megakaryocytes* break into fragments, which are the platelets. Their job is to begin the clotting process and plug up injured blood vessels. Platelets, also called *thrombocytes* (*thrombus* means "clot"), have a short life span — they live only about ten days.

## Putting up a good fight: White blood cells

White blood cells (WBCs) are derived from the same type of hematopoietic stem cells as RBCs. However, they take different paths early in the process of differentiation. The WBCs, also called *leukocytes*, leave the red bone marrow and enter into circulation in their mature form. (Flip to Chapter 13 for more on WBCs and immunity.)

### SQUEEZING BLOOD FROM A BONE

The process that forms blood cells is called *hematopoiesis*. In children, hematopoiesis occurs in the red bone marrow of the long bones such as the femur and tibia. In adults, it occurs mainly in the red bone marrow of the pelvis, cranium, vertebrae, and sternum. Special cells called *hematopoietic stem cells* divide and differentiate until they become specific blood cells. Turn to Chapter 3 for more about stem cells and cellular differentiation.

In brief, the process goes like this: A hematopoietic stem cell called a *hemocytoblast,* produces two lines of blood cell development. The *myeloid line* develops from the *myeloid stem cells* to form *erythrocytes, megakaryocytes,* and all *leukocytes* except lymphocytes. The *lymphoid line* develops from the *lymphoid stem cell* to form the *lymphocytes.* An *erythroblast* matures into an erythrocyte (RBC).

A *megakaryoblast* matures into a megakaryocyte, which fragments into platelets.

A myeloid stem cell can differentiate into any one of four types of WBCs: *basophil, eosinophil, neutrophil,* or *monocyte.*

After blood cells have matured in the red bone marrow, they enter circulation, which transports them around the body.

How do the hematopoietic stem cells "know" which kind of cell to "become"? The answer appears to be that they don't. All the main types of cells are produced at random. Then, factors in the hematopoietic microenvironment (the red bone marrow) cause some cells to "choose to die," a cellular process known as *apoptosis*. This is how the bone marrow can regulate the populations of the different types of cells.

# Looking at Your Blood Vessels

Your blood vessels comprise a network of channels through which your blood flows. But the vessels aren't passive tubes. Instead, they're very active organs that, when functioning properly, assist the heart in circulating the blood and influence the blood's constitution. The innermost layer of the heart and of all vessels is continuous — it's one very convoluted sheet of epithelium.

The vessels that take blood away from the heart are *arteries* and *arterioles*. The vessels that bring blood toward the heart are *veins* and *venules*. Generally, arteries have veins of the same size running right alongside or near them, and they often have similar names (turn to the "Arterial Components of the Cardiovascular System" color plate in the middle of the book for a diagram of the major arteries). The arterial vessels (arteries and arterioles) decrease in diameter as they spread throughout the body. Eventually, they end in the *capillaries*, the tiny vessels that connect the arterial and venous systems. The venous vessels become increasingly larger as they converge on the heart. The smaller venules carry deoxygenated blood from a capillary to a vein, and the larger veins carry the deoxygenated blood from the venules back to the heart.

## Starting with the arteries

Your arteries form a branching network of vessels, with the main trunk, called the *aorta*, arising from the left ventricle and splitting immediately into the *brachiocephalic trunk*, *left common carotid artery*, and *left subclavian artery*, which serve the head and upper limbs. The *descending aorta* — which serves the thoracic organs, abdominal organs, and lower limbs — gives off several branches. The first are the *mesenteric arteries*, the main arteries of the digestive tract. It also gives off two *renal arteries*, which supply the kidneys; and the *common iliac arteries*, which supply the pelvis and lower limbs. The arteries branch into smaller and smaller vessels, becoming arterioles that end in capillary vessels.

Although an artery looks like a simple tube, the anatomy is complex. Arteries are made of three concentric layers of tissue around a space, called the *lumen*, where the blood flows (see Figure 9-1). The outside layer, the *tunica externa*, is a thick layer of connective tissue that supports the vessel and protects the inner layers from harm. The larger the artery, the thicker the connective tissue supporting it.

The next layer in, the *tunica media*, is a thick wall of smooth muscle and elastic tissue. This layer expands and contracts with every pulse wave (heartbeat), and sometimes it gives a little squeeze to increase the pressure and force the blood through. This layer controls *vasoconstriction* (decreasing diameter) and *vasodilation* (increasing diameter) of the vessels.

Arterial lumen

Endothelium — ⎤
Basal lamina — ⎬ Tunica intima
Internal elastic ⎦ (Innermost layer)
membrane

Smooth muscle — ⎤ Tunica media
External elastic ⎬ (Middle layer)
lamina ⎦

Tunica adventitia
(externa)
(Outermost layer)

Vasa vasorum
(Small blood
vessels that
supply large
blood vessels)

K. BORN

**FIGURE 9-1:**
The anatomy of
an artery.

*Illustration by Kathryn Born, MA*

The inner layer, the *tunica interna,* is a single-cell thick layer of simple squamous epithelium that lines the lumen and is continuous through all the organs of the cardiovascular system. The *vascular endothelium,* as this tissue is also called, is very active metabolically, releasing into the blood various substances that influence blood circulation and vascular health. It also specializes in transporting oxygen, nutrients, and other substances from the blood flowing in the lumen to the smooth muscle fibers of the tunica media.

## Cruising through the capillaries

After passing from the arteries and the arterioles, blood enters the capillaries, which lie between larger blood vessels in *capillary beds.* A capillary bed forms a bridge between the arterioles and the venules.

Your capillaries are your smallest vessels: only the single-cell thick epithelial layer surrounds the lumen. The *precapillary sphincters* of the *metarterioles* (connectors of arteriole to capillary) can tighten or relax to control blood flow into the capillary bed.

Lacking the structure and complexity of arteries and veins, capillaries rely on simple diffusion to get their job done: moving oxygen and nutrients from the blood to the tissue and waste materials from the cells of the tissue to the blood. Molecules that are unable to diffuse exit the capillaries via filtration, driven out by the pressure applied on the capillary wall by the blood. *Capillary exchange* is the term used to describe these processes (see Figure 9-2).

Capillaries

Artery

Arteriole

Blood flow

Vein

Venule

Interstitial fluid moves in

Nucleus of endothelial cell

Plasma moves out

Blood flow

**FIGURE 9-2:** Capillary exchange.

*Illustration by Kathryn Born, MA*

The capillaries come into close contact with the cells of your tissues. At the end near the arteriole, oxygen diffuses out of the red blood cells and nutrient molecules out of the plasma, across the capillary membrane, and directly into the tissue fluid. The oxygen and nutrients dissolved in the tissue fluid diffuse across the membrane of the adjacent cells. (See Chapter 3 for more on diffusion.)

At the venule end, the carbon dioxide and other waste materials diffuse out of the tissue fluid and across the capillary membrane into the blood. Then, the blood continues on through the venous system, and those waste materials are deposited in the proper locations on their way out of the body. The carbon dioxide diffuses out of the bloodstream in the lungs, so it can be exhaled and thus removed from the body, while other metabolic waste is filtered through the kidneys.

Capillary beds are everywhere in your body, which is why you bleed anywhere that you even slightly cut your skin.

REMEMBER

Besides aiding in the exchange of gases and nutrients throughout your body, capillaries serve two other important functions:

>> **Thermoregulation:** Precapillary sphincters tighten when you're in a cold environment to prevent heat loss from the blood at the skin surface. Blood is then shunted from an arteriole directly to a venule through a nearby *arteriovenous shunt*. When you're in a warm environment or you're producing heat through exertion, the precapillary sphincters relax, opening the capillary bed to blood flow and dispersing heat.

>> **Blood pressure regulation:** When blood pressure (blood volume) is low, the hormones that regulate blood pressure stimulate the precapillary sphincters to tighten, temporarily reducing the total volume of the blood vessel system and, thus, raising the pressure. When blood pressure is high, the hormones stimulate the sphincters to relax, increasing overall system volume and reducing pressure. These hormones have the same effect on the larger vessels.

## Visiting the veins

Small veins converge into larger veins, all merging in the *inferior vena cava* and *superior vena cava*, the largest vessels in the venous system. These major veins return blood from below and above the heart, respectively. The inferior vena cava lies to the right of, and more or less parallel to, the descending aorta. The superior vena cava lies to the right of, and more or less parallel to, the aorta.

In the lower body, the *internal iliac veins*, which return blood from the pelvic organs, and the *external iliac veins*, which return blood from the lower extremities, converge into the *common iliac veins*. The *renal veins* return blood from the kidneys. Both these major veins flow into the inferior vena cava.

Blood from the digestive tract travels in the *hepatic portal vein* to the liver. Specialized cells in the liver move glucose molecules from the blood into storage. Phagocytic cells in the liver destroy bacterial cells that make it through the digestive process and remove toxins and other foreign material from the blood. Blood exits the liver through the *hepatic veins*, which flow into the inferior vena cava. The inferior vena cava empties into the right atrium.

Deoxygenated blood from the head and upper extremities drains into the *brachiocephalic veins*. The veins of the upper extremity — the *ulnar veins*, *radial veins*, and *subclavian veins* — also drain into the brachiocephalic veins. The *jugular veins* of the head and neck also drain into the brachiocephalic veins, which connect to the superior vena cava, which enters the right atrium.

After the blood from the right atrium has been pumped into the right ventricle, it's pumped into the lungs, where the blood is oxygenated, and then it flows back to the heart in the *pulmonary veins*, the only veins that carry oxygenated (red) blood.

Veins have a similar anatomy to arteries, although they tend to be wider and their walls thinner and less elastic. The tunica interna of a vein is also part of the continuous endothelial layer that lines the whole network. The tunica media has a layer of elastic tissue and smooth muscle, but this layer is much thinner in a vein than in an artery. The veins have virtually no blood pressure, so they don't need a thick muscle layer to vary the vessel diameter or withstand fluid pressure. The outermost tunica externa is the thickest layer of a vein.

Because veins don't have a thick muscle layer to push blood through them, they depend on contraction of skeletal muscles to move blood back to the heart. As you move your arms, legs, and torso, your muscles contract, and those movements "massage" the blood through your veins. The blood moves through the vein a little bit at a time. The larger veins have valves that keep blood from flowing backward. The valves open in the direction that the blood is moving and then shut after the blood passes through to keep the blood heading toward the heart.

# Cardiac Anatomy

The *cardiovascular system* (formerly known as the circulatory system) consists of the heart and the blood vessels. The heart's pumping action squeezes blood out of the heart, and the pressure it generates forces the blood through the blood vessels. The autonomic nervous system controls the rate of the heartbeat.

## Sizing up the heart's structure

The heart is shaped like a cone and is about the size of your fist in width and about two fists in length. It lies between your lungs, just behind (posterior to) your sternum, and the tip (apex) of the cone points down and to the left (see Figure 9-3). In most individuals, the heart is situated slightly to the left of center in the chest.

The four hollow spaces of the heart are called its *chambers.* You can check them out in the "Heart" color plate in the middle of the book. The heart is divided anatomically and functionally into left and right sides. Each side has one *atrium* and one *ventricle*, each with a separate function. A thin layer called the *interatrial septum* separates the atria; the *interventricular septum* separates the ventricles and is thick and muscular. The contraction of the heart muscle pumps blood into and out of all four chambers in a rhythmic pattern.

**FIGURE 9-3:**
A typical healthy heart.

Aorta

Pulmonary artery

Superior vena cava

Left atrium

Left coronary artery

Great cardiac vein

Left circumflex artery

Right coronary artery

Left anterior descending artery

Left ventricle

Right atrium

Anterior cardiac veins

Apex

Right ventricle

*Illustration by Kathryn Born, MA*

Between the chambers are *valves* that allow measured quantities of blood to flow into the chambers and keep blood flowing in the right direction. The valves' names tell you their anatomical location or characteristics. The two *atrioven-tricular* (AV) *valves* lie between the atrium and the ventricle on each side; the *bicuspid valve* (BV), often called the *mitral valve,* has two flaps; and the *tricuspid valve* (TV) has three flaps. The *semilunar valves* (SV) are shaped like half-moons. In Figure 9-4, the arrows show the direction of blood flow through the chambers of the heart.

During one heartbeat, blood fills the right atrium from the vena cavae. It then passes through the tricuspid valve, filling the right ventricle. It's then pumped through the pulmonary valve into the pulmonary artery. At the same time, the left atrium is being filled from the pulmonary veins. Blood then moves through the mitral valve into the left ventricle. It exits through the aortic valve into the aorta.

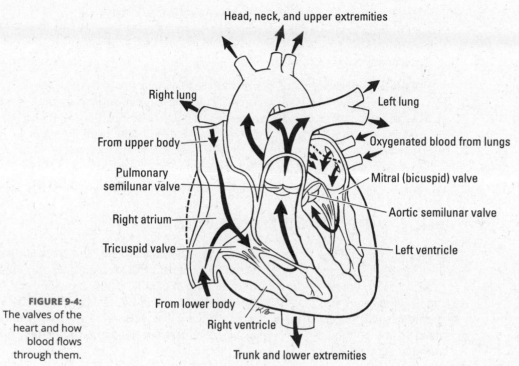

Head, neck, and upper extremities

Right lung

Left lung

From upper body

Oxygenated blood from lungs

Pulmonary semilunar valve

Mitral (bicuspid) valve

Right atrium

Aortic semilunar valve

Tricuspid valve

Left ventricle

**FIGURE 9-4:**
The valves of the heart and how blood flows through them.

From lower body

Right ventricle

Trunk and lower extremities

*Illustration by Kathryn Born, MA*

# Examining the heart's tissues

The tissues of the heart perform the functions required to keep the double-pump working strongly and steadily. Like other hollow organs, the heart is made up of layers of endothelial and connective tissue.

>> **Endocardium:** A layer of endothelial tissue that lines the inside of the chambers. This layer is continuous with the vascular endothelium, which we talk about in more detail in the "Starting with the arteries" section.

>> **Myocardium:** The thick, muscular layer of your heart. The myocardium (literally, "heart muscle") is composed of cardiac muscle fibers that contract in a coordinated way to pump the blood out of the heart and into the aorta with enough force to carry it through the arterial system and out into the capillaries.

>> **Epicardium:** The visceral layer of the *pericardium.* It's a conical sac of fibrous tissue that closely envelopes the myocardium and surrounds the roots of the major blood vessels. The epicardium secretes *pericardial fluid* into the *pericardial cavity,* which lubricates the tissues as the heart beats.

>> **Pericardial cavity:** A fluid-filled space between the epicardium and the parietal layer of the serous pericardium. The fluid reduces friction between the pericardial membranes.

>> **Parietal pericardium:** A serous membrane that's attached to the outermost layer of the heart, the *fibrous pericardium.* The fibrous pericardium is a thick, white sheet of fibrous connective tissue that anchors the heart and the major blood vessels, including the aorta, to the sternum and diaphragm. (Your heart is not just floating in your chest.) The parietal pericardium also secretes pericardial fluid into the pericardial cavity.

## Supplying blood to the heart

Blood flows in and out of the heart every second of your life, and some of that blood needs to supply the cells of the heart itself with oxygen and nutrients. Unfortunately, the heart cannot access it from the blood filling its chambers — capillary exchange still must take place in the myocardium. The coronary arteries supply oxygenated blood to the heart, while cardiac veins return deoxygenated blood to the pulmonary circulation loop (for the two loops of circulation, flip ahead to Figure 9-8).

>> **Coronary arteries:** Two large *coronary arteries* and their many branches supply blood to the heart. These large arteries branch off the aorta and carry blood to the left and right sides. They're called the left and right coronary arteries because they sit atop and encircle the heart, looking like a crown (see Figure 9-5). The *right coronary artery* and its two major branches, the *marginal artery* and the *posterior interventricular artery,* primarily supply the right atrium and ventricle with oxygenated blood and nutrients. The *left coronary artery* and its two branches, the *anterior descending coronary artery* and the *left circumflex coronary artery,* primarily supply the left atrium and ventricle with oxygenated blood and nutrients.

>> **Cardiac veins:** The cardiac venous system is similarly branched, often lying alongside the coronary arteries and their branches. Like everything about the heart, the cardiac venous system is composed of two parts, left and right. The *left cardiac venous system* receives deoxygenated blood from most of the superficial veins of the heart. The *right cardiac venous system* is composed of veins that originate on the anterior and lateral surfaces. The cardiac veins merge into the *coronary sinus,* which drains blood into the right atrium. The smallest vessels (the *venae cordis minimae*) drain the myocardium directly into the chambers.

**FIGURE 9-5:**
The coronary arteries.

Aorta

Left coronary artery

Left circumflex artery

Right coronary artery

Left anterior descending artery

*Illustration by Kathryn Born, MA*

# Cardiac Cycle

In order to keep our blood flowing smoothly and consistently through our bodies, the contractions of the heart must be intricately timed. Both atria must contract at the same time, followed by both ventricles shortly thereafter, creating one complete heartbeat. The *cardiac cycle* is defined as the series of events required to generate a single heartbeat. This section explores how the heart's electrical system triggers contractions and how those contractions generate pressure changes that push the blood along its path.

## Generating electricity

Certain structures of the heart, together called the *cardiac conduction system*, specialize in initiating and conducting the electrical impulses that induce your heartbeat, keeping it regular and strong in every part of the organ as it moves the blood through. The series of events that generate a single heartbeat is known as the *cardiac cycle.*

Throughout the *myocardium* (muscle wall of the heart) are specialized myocardial fibers (cells). Instead of being designed to contract, these fibers only carry impulse. These conducting fibers branch throughout their respective chambers so that the stimulus reaches the contracting fibers at the same time, creating the *syncytium*. The contracting fibers work in the same manner as the skeletal muscle fibers (see Chapter 6).

## Anatomy of cardiac conduction

Five structures comprise the cardiac conduction system (see Figure 9-6):

Atrioventricular bundle (bundle of His)

Left and right bundle branches

Sinoatrial node

Atrioventricular node

Purkinje fibers

*Illustration by Kathryn Born, MA*

**FIGURE 9-6:**
The conduction system of the heart.

>> **Sinoatrial (SA) node:** A small knot of cardiac muscle-like tissue (the conducting cells look like cardiac muscle cells, but they've lost the ability to contract) located on the back wall of the right atrium, near where the superior vena cava enters the heart. Once nervous input establishes the rhythm, the SA node becomes self-stimulating, which is why it's considered the heart's *pacemaker*. It does not require stimulus from a neuron to tell it to contract every time — only to tell it to speed up or slow down.

>> **Arterioventricular (AV) node:** A similar small mass of tissue located in the right atrium but near the septum that divides the right and left atria from the ventricles. Its function is to relay the impulses it receives from the SA node to the next part of the conduction system.

>> **Atrioventricular (AV) bundle (bundle of His):** A bundle of fibers that extends from the AV node into the *interventricular septum* (which divides the heart into right and left sides). The AV bundle transmits cardiac impulses.

>> **Left and right bundle branches:** Where the septum widens, the AV bundle splits into the right and left bundle branches, each extending down toward the apex and then up along the outside of the ventricles. These bundles carry cardiac impulses.

>> **Purkinje fibers:** At the ends of the AV bundle branches are the Purkinje fibers, which deliver the impulse up and around to the myocardial fibers, causing the ventricles to contract.

**TECHNICAL STUFF**

The word *node* is used in many different contexts, and not only in anatomy. In general, it describes a local connecting point of component parts, literal and figurative. In cardiac anatomy, a *node* is a specialized type of tissue that looks like muscle and generates electrical impulses like nervous tissue.

## The sequence of events

The cardiac conduction system is responsible for keeping the cardiac cycle going. If the cardiac cycle stops for too long, you experience serious consequences (see the "Cardiac disorders" section later in the chapter). It is also responsible for the timing of the cycle. A single heartbeat actually consists of two contractions. First, both the right and left atria contract together, pushing blood down into the ventricles. Moments later, the right and left ventricles contract together pushing the blood out of their respective arteries.

Here's how the cycle goes:

1. **The electrical impulse is initiated in the SA node.**

   The conducting fibers spread the impulse throughout the atrial syncytium. The right and left atria contract simultaneously, pumping blood into the right and left ventricles, respectively.

2. **The impulse passes to the AV node, which sends it to the AV bundle.**

   Some of the fibers connected to the SA node carry the impulse to the AV node. These fibers are narrower, providing the delay between the atria and ventricles contracting.

3. **The impulse passes into the right and left bundle branches and ulti-mately into the Purkinje fibers, causing the ventricles to contract.**

   By the time the ventricles begin to contract, the atria are relaxing. The impulse is carried to the apex first and then back up for two reasons:

   ● To further ensure the delay before the ventricles contract

   ● To allow the ventricles to contract in a pattern

   The purkinje fibers weave into the myocardium of the ventricles in a spiraling pattern. This causes the ventricular syncytium (walls of both ventricles) to contract in an upward, twisting motion.

4. **The ventricles relax.**

   All chambers remain relaxed until the SA node generates the next impulse, starting the cycle over again.

## Moving blood through the heart

In order for the blood to move on its well-timed path through the heart, we must coordinate the valves opening and closing just as we do the contractions. It's rea-sonable to conclude that the contractions are the cause of the valves opening and closing, but it's actually more complicated than that.

At the start of the cardiac cycle, all the chambers are relaxed and all the valves are closed. When a chamber is relaxed, it is said to be in *diastole*; the contraction phase is called *systole.* The sequence of events is as follows:

1. **Blood begins filling the atria, the pressure on the AV vale pushes it open.**

   Both chambers are still in diastole.

2. **Blood flows into the ventricle to equalize the pressure between the two chambers.**

   Blood stops flowing when about 70 percent of the volume to be pumped out is in the ventricle.

3. **The atria contract, pushing the remaining blood into the ventricles.**

4. **The atria relax, the ventricles contract, and the AV valves close.**

   This drop in pressure of the atria creates a vacuum, which pulls the flaps of the AV valves closed. The *papillary muscles* contract with the walls, pulling on the *chordae tendinae* to ensure the valves stay closed (see the "Heart" color plate for the location of these structures).

5. **Ventricles continue contracting.**

   The atria are relaxed, all valves are closed.

6. **The ventricular systole increases the pressure inside the ventricles, pushing open the semilunar valves.**

   Blood exits into the pulmonary artery and aorta.

7. **The ventricles relax, decreasing the pressure, which pulls the semilunar valves closed.**

   The drop in pressure causes blood to start to flow backward, but it catches in the cups of the valves, snapping them shut.

REMEMBER

The movement of blood through the heart is more a result of pressure changes than contractions directly. The intricate timing provided by the conduction system triggers the contractions, which provide the pressure changes that both open the valves so the blood may exit and close them to prevent backward flow.

## The heartbeat

If you're a fan of medical dramas, you're no doubt familiar with the heart rate monitor — those squiggly lines moving across the screen. The implication is that those waves correspond to the contraction of the chambers. Actually, this isn't quite the case.

The heart's electrical signals can be measured and recorded digitally to produce an image called an *electrocardiogram* (ECG or EKG). Three major waves appear on the EKG, each showing the electrical signals as they move through the heart (see Figure 9-7). The first wave, called the *P wave*, records the spread of impulse through the right and left atria. The second and largest wave, the *QRS wave*, shows the spread of impulse through the right and left ventricles. The third wave, the *T wave*, records the recovery of the ventricles.

The conducting fibers spread impulse just as neurons do (see Chapter 7). The spikes on the EKG correspond to the movement of ions, measurable by electrodes on the skin because body fluids conduct this electricity. As the conducting fibers depolarize, the EKG line moves up and back down. The same happens when they repolarize (because the flow of $K^+$ ions out is measurable electricity just as the flow of $Na^+$ ions in). Because the repolarization of the atria occurs during the QRS, you don't see a wave for it.

REMEMBER

The waves on an EKG show the electrical conduction in the heart, not the contraction. The atria would be contracting in the period of time between the end of the P wave and the start of the QRS (roughly). The ventricles are contracting from the end of the QRS to the T wave.

© John Wiley & Sons, Inc.

**FIGURE 9-7:**
The typical EKG
pattern.

Problems with signal conduction due to disease or abnormalities of the conduct-ing system can occur anyplace along the heart's conduction pathway. Abnormally conducted signals resulting in irregular heartbeat are called *arrhythmias*.

# Physiology of Circulation

You do it 100,000 times every day. Waking or sleeping, from a moment early in fetal development until the moment you die, the beat goes on. It may speed up slightly when you're working hard physically or you're excited or stressed, and some people can slow theirs down with meditation techniques. But for most peo-ple, it just plugs along, the same thing over and over.

What is it? The heartbeat, of course. The beating heart pushes blood around a double-circuit — out through your arteries, ultimately into your capillary beds, then across your capillary beds and into your veins, and back to the heart. The blood passes through the heart to the lungs and then back to the heart and out through the arteries again. Each complete double-circuit takes less than one minute.

## On the beating path: The circuits of blood through the heart and body

**REMEMBER**

The heart is a double-pump, so it has two circuits: heart to lungs and back to heart, and heart to body and back to heart. These are called the *pulmonary circula-tion path* and the *systemic circulation path*, respectively (see Figure 9-8). Every drop of your blood travels around the double-circuit about once per minute.

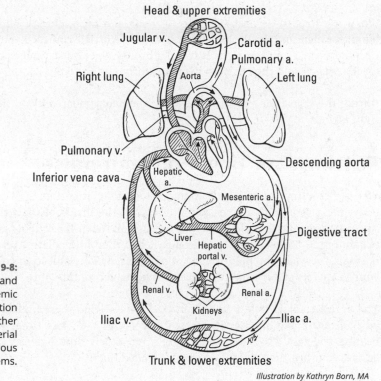

Head & upper extremities

Jugular v.

Carotid a.

Pulmonary a.

Right lung

Aorta

Left lung

Pulmonary v.

Descending aorta

Inferior vena cava

Hepatic a.

Mesenteric a.

Digestive tract

Liver

Hepatic portal v.

Renal v.

Renal a.

Kidneys

Iliac v.

Iliac a.

Trunk & lower extremities

**FIGURE 9-8:**
Pulmonary and
systemic
circulation
working together
through arterial
and venous
systems.

*Illustration by Kathryn Born, MA*

## Pulmonary circulation

Deoxygenated blood enters the heart's right atrium from the largest veins in the body, the superior vena cava and the inferior vena cava. When the SA node initiates the cardiac conduction cycle, the right atrium contracts, pumping the blood into the right ventricle.

When the impulse is passed to the AV node and on to the AV bundle, the right bundle branch, and the Purkinje fibers, the right ventricle contracts, pumping blood into the pulmonary arteries, which take it to the lungs for gas exchange.

During the relaxation phase of the atria, the newly oxygenated blood flows into the left atrium.

**TECHNICAL
STUFF**

The pulmonary arteries are the only arteries that carry deoxygenated blood, and the pulmonary veins are the only veins that carry oxygenated blood.

## Systemic circulation

When the SA node initiates the cardiac conduction cycle, the left atrium contracts, pumping the oxygenated blood into the right ventricle.

When the impulse is passed to the AV node and on to the AV bundle, the left bundle branch, and the Purkinje fibers, the left ventricle contracts, pumping blood into the aorta. From the aorta, it travels through the arteries and arterioles to the capillary beds, and then back to the heart through the veins.

During the relaxation phase of the atria, the deoxygenated blood flows into the right atrium.

## Putting your finger on your pulse

You can feel the rhythmic pulsation of blood flow at certain spots around your body, most commonly on the radial artery on the inside of the wrist or on the carotid artery of the neck. What you feel as you touch these spots is your artery expanding as the blood rushes through it and then immediately returning to its normal size when the bulge of blood has passed. This pulsation corresponds to your heart rate and is usually recorded as beats per minute.

**TECHNICAL STUFF**

The entire cardiac cycle takes about 0.86 of a second, based on the average of 70 heartbeats per minute. If your cardiac cycles take less time, your heart is beating too fast (*tachycardia*); if there's too much time between your cardiac cycles, your heart is beating too slowly (*bradycardia*).

## Going up, going down, holding steady: Blood pressure

*Blood pressure* is a term used to describe the force of blood pushing against the wall of an artery. It's measured in millimeters of mercury at both the highest point (*systole,* when the heart is contracted) and the lowest point (*diastole,* when the heart is relaxed) in the cardiac cycle. The systole pressure is always higher than the diastole. The higher the systolic and diastolic values, the more pressure there is on the walls of the arteries. Two factors affect the blood pressure: the *cardiac output,* which is the amount (volume) of blood the heart pumps out per unit of time, and the *peripheral resistance,* a measure of the diameter and elasticity of the vessel walls.

The cardiac output is determined by the heart rate and the blood volume put out from a ventricle during one beat (reported in L/min.). When either of these rises, the blood pressure rises. Heart rate is increased by physical exertion, the release of epinephrine (a hormone), and other factors. The blood volume is influenced by the action of ADH (antidiuretic hormone) and other mechanisms in the kidneys to control the amount of water that's removed from the urine and restored to the blood.

The arteries' diameter changes locally and continuously. The pressure of the pulse wave increases pressure on the vascular endothelium, inducing it to release molecules, mainly nitrous oxide (NO), that induce relaxation in the tunica media. The endothelium's ability to respond to the pulse-wave pressure is extremely important to vascular health. Resistance in the arteries to expansion as the blood rushes through raises the blood pressure.

As part of homeostasis, receptors in the arteries, called *baroreceptors,* measure blood pressure. If the blood pressure is above the normal range, the brain sends out impulses to cause responses that decrease the heart rate and dilate the arterioles, both of which decrease blood pressure.

## Not going with the flow

Not the least amazing thing about blood is its ability to stop flowing. The term for this is *hemostasis* (literally, "blood stopping," and not to be confused with *homeostasis*). Hemostasis is the reason why you didn't bleed to death the first time you cut yourself. When vessels are cut, blood flows, but only for a moment, just about long enough to clean the cut. As you watch, the blood stops flowing, and a plug, called a *clot,* forms. Within a day or so, the clot has dried and hardened into a scaly scab. Eventually, the scab falls off, revealing fresh new skin.

A blood clot consists of a plug of platelets enmeshed in a network of insoluble fibrin molecules. The *clotting cascade* is a physiological pathway that involves numerous components in the blood itself interacting to create a barrier to the flow. Each step in the process triggers the next so the whole mechanism is an example of positive feedback (see Chapter 2). As soon as a blood vessel is injured, a signal is sent from the vascular endothelium to the platelets, also called *thrombocytes* (*thrombo* means "clot"), summoning them to the injury site. The platelets become sticky, and they start to adhere together. Proteins in the blood plasma, called *coagulation factors* or *clotting factors,* respond in a complex cascade to form fibrin strands, which strengthen the platelet plug. Within minutes, your blood is safe in your body, where it belongs.

A blood clot in the right place is something to be grateful for. But blood tends to start clotting whenever it's not flowing freely, and this tendency can cause problems in the peripheral vessels (the arteries and veins of the legs). Clots also form on the inner wall of blood vessels when the endothelium is injured by disturbances in blood flow (turbulence) or by free radicals in the blood. These tiny clots adhere to the wall, further disturbing flow and causing more turbulence and more injury. Atherosclerotic plaque may begin to form around the clot (see the next section for a discussion of atherosclerosis). Worst of all, perhaps, is when the clot, with the plaque attached, breaks off from the vessel wall and floats free (sort of) in the bloodstream. Sooner or later, this *embolus* lodges somewhere in a vessel, sometimes with sudden and fatal consequences such as a stroke.

# Pathophysiology of the Cardiovascular System

The cardiovascular system endures constant mechanical stress (sheer stress, from the pressure of fluid flow), physical blockages large and small, and other physical and chemical assaults. The blood is a natural target for pathogens and parasites of all kinds.

## Cardiac disorders

The heart muscle is subject to malfunctions of electrical and chemical signaling that disrupt its all-important rhythm.

### Arrhythmia

An *arrhythmia* is any disorder of your heart rate or rhythm, such as the heart beating too quickly (faster than 100 beats per minute, called *tachycardia*), too slowly (fewer than 50 beats per minute, called *bradycardia*), or in an irregular pattern. *Fibrillations* are another type of arrhythmia where the walls of the chambers contract spasmodically. The syncytium does not contract all at once, affecting the amount of blood that continues on its path. *Atrial fibrillation* (known as afib) is generally milder because the ventricle still receive blood. *Ventricular fibrillation* (known as vfib) results in a stoppage of blood flow or *cardiac arrest.*

Arrhythmia, also called *dysrhythmia,* can originate as a disorder of any of the structures of the cardiac conduction system and any part of the heart can be affected. In general, arrhythmias that start in the ventricles are more serious than those that start in the atria. Since 70 percent of the cardiac output is in the ventricles before the atria contract, blood will keep flowing if the ventricles still contract properly.

Transient arrhythmia can be brought on by many factors, including caffeine, vigorous exercise, and emotional stress. Some arrhythmias are chronic conditions, treated with an implantable medical device called a *pacemaker,* which initiates the impulse in the area that's not receiving it. The most serious of all cardiac arrhythmias is vfib. Treatment often involves a *defibrillator,* which shocks the patient's heart to reestablish the normal electrical flow.

### Myocardial infarction

*Myocardial infarction* (or "heart attack") is the irreversible damage of myocardial tissue caused by a blockage of blood flow to the myocardium — the occlusion of a coronary artery. The myocardial fibers supplied by that artery become oxygen-depleted and die or become irreversibly damaged (infarction). Depending on the size of the infarct and its location, the severity of the follow-on effects on cardiac rhythm and the integrity of the myocardium itself can range from relatively mild to fatal.

Occlusion of a coronary artery may be from a blood clot (coronary thrombosis) following the rupture of an atherosclerotic plaque on the vascular endothelium.

# Vascular disorders

Vascular disorders are malfunctions of the blood vessels. The following examples are disorders of the arteries. The veins, particularly the deep veins of the legs, are subject to damage and malfunction, too.

## Hypertension and atherosclerosis

*Hypertension,* more commonly called "high blood pressure," means that blood is pushing too hard on the walls of the arteries. This can happen because cardiac output is high or because the arteries themselves have lost the ability to flex in response to the pulse-wave of blood.

As a fast-flowing stream damages the stream bank, the pressure of the blood flow exerts *shear stress* on the thin and vulnerable vascular endothelium. Platelets rush to the injury site and initiate the clotting cascade. A clot *(thrombus)* may form on the artery wall, intruding into the lumen and possibly blocking the artery.

The cumulative damage to arteries causes many problems: *ischemia* (reduced blood flow to tissues and organs); *angina* (sudden pain as blood squeezes through arteries); worsened shear stress because of turbulence in the blood flow; and *embolus,* a clot dislodged from the artery wall, which can be fatal.

Microinjuries in the artery walls lead to the formation of *arterial plaque.* Fats in the blood build up on the epithelial lining. Then, fibrous plaques begin to form on the fatty deposits. Next, calcium deposits become enmeshed in the fibrous plaques. *Atherosclerosis* is the narrowing and stiffening of the arteries due to arterial plaque. Atherosclerosis increases the risk of coronary artery disease (CAD) and myocardial infarction (MI).

## Stroke

A *stroke* is damage to the brain caused by *ischemia* (reduced blood flow) or *hemor-ragia* (bleeding). Ischemia results from atherosclerosis in the arteries that supply the brain — the internal carotid arteries and the cerebral arteries. Bleeding may be caused by a ruptured vessel in the brain (an *aneurysm*). Whether ischemic or hemorragic, some brain tissue dies or is irreversibly damaged.

Disability following a stroke may be mild. The patient may even make a complete recovery, as other parts of the brain may start "filling in for" the damaged areas by acquiring new capabilities. Or damage may be severely disabling, both physically and mentally. Frequently, people who experience stroke ultimately experience several more strokes of increased severity. Stroke is a common cause of death among the elderly.

When a stroke is disabling, the specific nature of the disability depends on which part of the brain is affected. Speech problems, from slight slurring to total *aphasia* (inability to speak), may result from damage to the speech control areas in the left frontal lobe. If the brain tissue that dies is in the center for vision, the stroke victim may become blind. Damage to motor-control areas cause problems ranging from *paresis* (impaired movement) in one eyelid to serious paralysis that affects one or both sides of the body.

# Blood disorders

Any disruption or change in the composition and chemistry of blood is a disorder. Here you see how such changes can disrupt the delivery of oxygen, needed by all cells. You also see how parasites can divert the resources of the blood for their own processes.

## Anemia

*Anemia*, the most common blood disorder, is a malfunction of RBCs resulting in low blood oxygen levels. The underlying cause may be related to any of the following:

>> A low number of RBCs, which, in turn, can have a number of underlying causes related to cell differentiation and maturation in the bone marrow, or some process whereby RBCs are lost, such as chronic internal bleeding (called *occult bleeding*) or *hemolytic anemia* — when RBCs are destroyed at a higher rate than they're produced.

>> A low quantity of hemoglobin in the RBCs, which may result from a deficiency of iron in the diet.

>> Impaired oxygen-binding capacity in the hemoglobin molecules. Several diseases called *hemoglobinopathies* have this effect.

## Sickle cell disease

*Sickle cell disease* (SCD) is a *hemoglobinopathy,* an inherited (genetic) disorder of the pathway for producing hemoglobin molecules. The abnormally shaped hemoglobin molecules that are produced have impaired oxygen-binding capacity. The RBCs containing abnormal hemoglobin have a life span even shorter than the life span for normal RBCs and often rupture in the blood. Free hemoglobin in the blood is toxic to the vascular endothelium's cells.

The abnormal hemoglobin also distorts the shape of the RBCs, which causes blockages in the blood flow, setting off a *syndrome* (a clinical term for a group of related signs and symptoms) called a *sickle cell crisis.* A number of factors or events can trigger a sickle cell crisis, such as cold temperature, high altitude, physical or emotional stress, infection, and low blood oxygen level *(hypoxia).* Blocked blood flow initiates a clotting cascade, causing pain and possibly necrosis (dead tissue) in the area, and damaging the artery wall. Damaged arteries leave the patient vulnerable to all the resulting disorders of atherosclerosis and hypertension; the prevalence of pulmonary hypertension, a very dangerous condition, is particularly high in the SCD patient population.

## Malaria

*Malaria* is a disease caused by infection with a single-celled, parasitic organism named *Plasmodium.* The disease is transmitted through the bite of a particular genus of mosquito. When the mosquito feeds on a host with malaria, the parasite makes its way into the mosquito's saliva. When it bites the next person, even a week later, the parasite enters the host's blood flow. Within the RBCs, the parasites multiply, periodically undergoing amplification cycles and breaking out en masse to invade fresh RBCs. During such cycles, the patient suffers waves of fever, chills, nausea, and malaise.

No one knows for sure how long malaria has been endemic in human populations, but it has been long enough for the human genome to evolve a genetic response in the form of altered hemoglobin molecules like those seen in sickle cell disease. Present in just the right amount, these abnormal hemoglobins can confer some protection from malaria.

Chapter **10**

# The Respiratory System: Breathing Life into Your Body

The air we inhale more than 20,000 times each day is mostly made of nitrogen, but it also contains the vital, irreplaceable gas that is oxygen. Without oxygen, our cells can't perform cellular respiration efficiently enough to generate the ATP they require (see Chapter 2). We do need nitrogen to build proteins, but we aren't equipped to obtain nitrogen from the air we breathe. Getting the oxygen into our blood is what the respiratory system is all about, so take a deep breath and prepare to find out all about it in the pages ahead.

## Functions of the Respiratory System

Your respiratory system manages the flow of air into and out of your body. It oversees a number of vital body functions, including the following:

» **Ventilating:** *Ventilation* — the act of breathing — isn't the same as *respiration* — the exchange of gases. You don't have to think about performing either task, but ventilation involves the physical movement of your diaphragm muscle, rib cage, and lungs to draw air in and push air out of the body. See the "Breathing: Everybody's Doing It" section later in this chapter.

>> **Exchanging gasses:** The job of the respiratory system is to supply oxygen and remove carbon dioxide from the blood as it circulates through the body. Chapter 9 discusses how the blood gets into and out of the lungs in the process of *pulmonary circulation.* This is sometimes called *external respiration* to distinguish it from *cellular respiration.* Gas exchange happens in the lungs, where the respiratory tissues and the circulatory tissues meet. See the "Gas exchange" section later in the chapter.

>> **Regulating blood pH:** Maintaining blood pH within the homeostatic range requires coordination between the respiratory and urinary systems, with help from the endocrine system and, of course, the cardiovascular system itself.

>> **Producing speech:** The ability of humans to consciously control breathing permits speech and singing.

# Nosing around Your Respiratory Anatomy

The *respiratory tract* is the path of air from the nose to the lungs. It's divided into two sections: the *upper* respiratory tract (from the beginning of the airway at the nostrils to the pharynx) and the *lower* respiratory tract (from the top of the trachea to the diaphragm). Turn to the "Respiratory System" color plate in the center of the book to get an idea of what the internal structures of the respiratory tract look like.

The respiratory tract is one of the places in the body where cells are replaced constantly throughout life (see Chapter 2).

## Nose

This is one time that it's okay to turn your nose up at your anatomy. Really. Point your nose up while looking in the mirror. (Yes, you have to put the book down.) See the two big openings? Those are your *nostrils,* and that's one of two places where air enters and exits your respiratory system. Now, see all those tiny hairs in your nostrils? Those little hairs serve a purpose. They trap dirt, dust particles, and bacteria. Okay. You can put your head down now. The rest of your respiratory parts are way inside of your body, so you can't see them in the mirror.

Just beyond your nostrils, the *nasal septum* separates your *nasal cavities.* Inside the nasal cavities, the three tiny bones of the *nasal conchae* provide more surface area inside the nose because they're rolled up (like conch shells). The cells of the *respiratory mucosa* that lines the inside of the nasal cavity have tiny *cilia* that move the dirt-laden mucus toward the outside of the nostrils.

The *lacrimal glands* secrete tears that flow across the eye's surface and drain through the openings in the corner of the eye *(lacrimal puncta),* into the *nasolacrimal ducts* and the nasal cavities. That's why your nose runs when you cry.

Your *sinuses* are air spaces in your skull that lighten the weight of your head. The sinuses open into the nasal cavities so they can receive air as you breathe, and, like the nasal cavities, the sinuses are lined with mucous membranes. The sinuses also contribute to the tone of your voice.

## Pharynx

Air passes through your *pharynx* on its way to your lungs. Along the way, it passes through and by some other important structures, like your larynx and tonsils.

Your pharynx is divided into three regions based on what structures open into it:

**TECHNICAL STUFF**

>> **Nasopharynx:** The top part of your throat where your nasal cavities drain. If you press your tongue to the roof of your mouth, you can feel your *hard palate.* This bony plate separates your mouth *(oral cavity)* from your nose *(nasal cavities).* If you move your tongue backward along the roof of your mouth, you reach a soft spot. This spot is the *soft palate.* Beyond the soft palate is where your nasal cavities drain into your throat, your *nasopharynx.* Your soft palate moves backward when you swallow so that the nasopharynx is blocked.

Normally, the soft palate blocking the nasopharynx keeps food from going up into your nose. But when you're laughing and eating or drinking at the same time, your soft palate gets confused. When you go to swallow, it starts to move back, but when you laugh suddenly, it thrusts forward, allowing whatever's in your mouth to flow up into your nasal cavities and immediately fly out of your nostrils to the delight of everyone around you.

>> **Oropharynx:** The middle part of your throat, frequently called "the back of the throat." It extends from the *uvula* to the level of the *hyoid bone.* It's the location of the *epiglottis,* a cartilage structure that guides materials passing though the mouth to the trachea or esophagus, as appropriate. It's why your food comes close to your windpipe but rarely goes in. We mention the special characteristics of the hyoid bone in Chapter 5. We describe the esophagus, part of the digestive tract, in Chapter 11.

# MUCUS: GROSS, BUT NECESSARY

Inquiring minds want to know: What is mucus? Where does it come from? What does it do?

*Mucus* (the adjective form is *mucous*) is a thick fluid made up of water, salts (electrolytes), glycoproteins called *mucins,* enzymes, epithelial cells, and immune system components like *immunoglobulins* (antibodies) and *leukocytes* (white blood cells). In mammals, including humans, mucus is found in the structures of the respiratory, digestive, reproductive, and urinary systems. Many other animals secrete mucus, including many invertebrates, notably snails and slugs.

Mucus is produced in the *mucous membranes,* or *mucosae* (singular, *mucosa*). These membranes have cells and glands that secrete mucus. Mucosae line the airway (nose, trachea, bronchus, and bronchioles), the entire digestive tract, the ureters and urinary bladder, and reproductive tracts of both males and females. The average human body produces about a liter of mucus per day.

In the respiratory system, mucus oozes in a continuous stream from the mucosa that lines the entire airway. Cilia on the epithelial cells of the bronchi and bronchioles constantly sweep this slow stream of mucus back up the airway toward the throat (pharynx). In the nasal passages, mucus warms and moistens the inhaled air before it reaches the delicate *alveoli.* Mucus coats the hairs that line the nasal passages, trapping dust and irritating particles in its sticky mass and expelling them back outside the body. Its immune system components destroy airborne pathogens. Any particles that escape the nasal hairs are engulfed farther down the airway. The continuous movement of the respiratory mucus layer toward the oropharynx helps prevent foreign objects from entering the lungs during breathing. Over-secretion of mucus is a common symptom in inflammatory respiratory ailments such as colds, the flu, respiratory allergies, asthma, and chronic bronchitis. Sneezing clears extra mucus from the upper respiratory tract while coughing clears the lower.

The digestive system is lined by mucosa for its entire length. Mucus serves to moisten and soften the digesting material and to lubricate its passage. A thick coat of mucus protects your stomach lining from the extremely acidic conditions inside. Mucus isn't digested in the intestinal tract, so mucus commonly appears in fecal matter.

In the female reproductive system, cervical mucus prevents infection. In the male reproductive system, the semen contains mucus. Both sexes secrete mucus-containing lubricating fluids during sexual intercourse. In other systems, mucus serves primarily to protect and lubricate surfaces.

>> **Laryngopharynx:** The lower part of your throat adjacent to your larynx. The *larynx* (or *voice box*) is triangular. At the apex of the triangle is *thyroid cartilage,* commonly known as your "Adam's apple." If you could look down your throat onto the top of your larynx, you'd see your *glottis,* the opening through which air passes. When you swallow, a flap of tissue called your *epiglottis* covers your glottis and blocks food from getting into your larynx.

Inside your larynx are the *vocal cords* — gathered mucous membranes that cover ligaments. Your vocal cords vibrate when air passes over them, producing sound waves. Pushing more air over them increases the vibration's amplitude, making the sound louder. When you tighten your vocal cords, the glottis narrows, and your voice has a higher pitch.

# Trachea

Your *trachea* (windpipe) is a tube that runs from your larynx to just above your lungs. Just behind your sternum, your trachea divides into two large branches called *primary bronchi* (singular, *bronchus*) that enter each lung.

The trachea and bronchi are made of epithelial tissue, as well as smooth muscle and cartilage, which allow the airways to constrict and expand.

# Lungs

Your lungs are large paired organs within your chest cavity on either side of your heart. They're spongy and, like the heart, are protected by the rib cage. The lungs sit on top of the *diaphragm,* a powerful muscle that's fixed to the lower ribs, sternum, and lumbar vertebrae. The heart sits in a depression between the lungs, called the *cardiac notch.*

The left lung is smaller than the right, to make room for the heart. Both lungs are separated into *lobes* (three on the right and two on the left). The lobes are further divided into segments and then into *lobules,* the smallest subdivision visible to the eye.

## Pleural sac

Each lung is completely enclosed in the *pleural sac.* The pleural sac is similar to the pericardial sac in that it's made up of two membranes, the *parietal* pleura, attached to the thoracic wall, and the *visceral* pleura, attached to the lung's surface, with the pleural cavity between them. The pleural cavity contains a lubricating fluid called the *intrapleural fluid.*

The adhesive force of the fluid interface between the parietal pleura and the visceral pleura connects the lungs to the chest wall. In other words, the two membranes contain molecules that really like to "hold hands." So when your chest rises as you inhale, the lungs expand with it and vice versa.

The intrapleural fluid completely surrounds the lungs. It keeps the pleural membranes moist and lubricated. Because of this fluid, the pleural cavity has a negative pressure (lower than atmospheric pressure), which keeps the lungs inflated.

### The bronchial tree

After the primary bronchus enters the lung on each side, it splits into secondary and tertiary branches called *bronchi.* The tertiary bronchi divide into smaller branches called *bronchioles,* which transport air to the lobules. At the end of the smallest bronchioles are little structures that look like clusters of grapes; these are the *alveolar sacs.* The individual "grapes" are *alveoli* (singular *alveolus*). This is where the lungs interact with the blood vessels for gas exchange to occur.

TECHNICAL
STUFF

The lungs have a very large reserve volume relative to the oxygen exchange requirements when at rest. As a result, it's possible to live many years with only one lung.

## Diaphragm

The *thoracic diaphragm* is a dome-shaped sheet of muscle separating the base of the lungs from the liver, and, on the left side, from the stomach and the spleen. The diaphragm pushes up beneath the lungs to control their contraction and expansion during ventilation. The motor fibers in the *phrenic nerves* signal to the diaphragm when to contract and relax. The diaphragm can also exert pressure on the abdominal cavity, helping the expulsion of vomit, feces, or urine.

# Breathing: Everybody's Doing It

Breathing is essential to life, and thankfully, your body does it automatically. Air is alternately pulled into (inhaled) and pushed out of (exhaled) the lungs because of changes in pressure. When you inhale, your chest cavity expands, as do your lungs. This decreases the pressure inside the lungs, causing the air to flow through the bronchial tree and into the alveoli. When you exhale, the volume of the thoracic cavity decreases, which increases the pressure inside the lungs, forcing some air out (see Chapter 16). In the following sections, we take a look at how your body breathes under different conditions.

# Normal breathing

When you're sleeping, sitting still, and doing normal activity, your breathing rate is 12 to 20 inhalation/exhalation cycles per minute. Normal breathing *(eupnea)* is involuntary, which is why you never really forget to breathe, even when you're sleeping. In many cases, breathing continues even during a coma. Impulses to the diaphragm come through a pair of spinal nerves, called the *phrenic nerves.* They initiate the regular, alternating contraction and relaxation of the diaphragm. The rhythm of the impulse is controlled by the autonomic system in the brain stem.

**TECHNICAL STUFF**

While your respiration rate is involuntarily controlled by the respiratory center — the pons and medulla oblongata of your brainstem — you can also voluntarily control your breathing. The diaphragm is still triggered to contract, but the order originated from the cortex of the brain instead.

*Inspiration* (breathing in) is the result of the diaphragm's contraction. Seems backward, doesn't it? When the diaphragm contracts, it pushes downward into the abdomen, creating more space in the thoracic cavity. The *intercostal muscles* (the muscles between the ribs) may also be stimulated to contract, helping to expand the lungs even more by pulling the ribs up and out. Air moves in at the top of the airway (through the nose or the mouth if necessary) and all the way down to take up the expanded space in the alveoli.

During normal breathing, *expiration* (breathing out) is a passive process requiring no energy or instruction. The brain simply stops sending impulses to the diaphragm, causing it to relax. As it settles back into position, the volume of the thoracic cavity decreases. Additionally, all the elastic tissue in the lungs recoils (think about stretching and then letting go of a rubber band). This increases the pressure in the lungs, forcing the air out through the respiratory tract (see Figure 10-1).

# Breathing under stress

Keep in mind as you read this section that "stress" just means that an extra physiological demand is being placed. Stress isn't necessarily negative — whether physical or emotional, stress can be painful or pleasurable, and is often both.

However much fun you are or are not having, stress increases metabolism. More oxygen is consumed, and more carbon dioxide is produced. An increase in carbon dioxide will decrease your pH, triggering *chemoreceptor cells* in the carotid arteries and aorta. Surprisingly, the amount of oxygen in your blood does little to influence your respiratory center.

Inspiration

Expiration

Air
Rib cage

Air
Rib cage

Rib cage
moves out
and up

Rib cage
moves down
and in

Diaphragm
moves down

Diaphragm
moves up

Lung pressure
decreases and
air rushes in

Lung pressure
increases and
air pushed out

Diaphragm

Diaphragm

**FIGURE 10-1:**
The inspiration
(inhalation) and
expiration
(exhalation)
process.

*Illustration by Kathryn Born, MA*

Once your respiratory center is alerted, both inspiration and expiration become active processes. Your breathing is deeper and more frequent. The intercostal muscles contract more forcefully, and the *pectoralis* and *sternocleidomastoid* may also assist in the process. This decreases the pressure even more and air rushes in. During exhalation, the abdominal muscles will contract, forcing the diaphragm up and pushing out more air (so that a greater volume can be brought in with the next breath). These processes restore homeostasis and support the elevated metabolism.

TECHNICAL
STUFF

The forcible exhalation involved in coughing or sneezing is aided by the sudden contraction of the abdominal muscles, raising the abdominal pressure. The rapid increase in pressure pushes the relaxed diaphragm up against the pleural cavity, forcing air out of the lungs.

## Controlled breathing

As far as has been determined, humans are the only animals who can bring their breathing under conscious control. Breathing also allows people to speak and sing, as well as moderate other physiological systems.

# TAKING MEASURED BREATHS

To determine whether you're breathing properly and your lungs are functioning well, physicians can measure the amount of air you inhale and exhale, as well as how much air remains in your lungs after you exhale and how much air you're capable of holding.

The *tidal volume* (TV) is the volume of air inhaled or exhaled in a single breath during normal, relaxed breathing. This amount is about half a liter.

Of course, if you breathe deeply, you can inhale and exhale more. The maximum volume of air that you can move through your lungs in one breath is your *vital capacity* (VC). This is determined by taking the deepest breath possible, and then measuring how much air can be forcefully exhaled afterward. By breathing really deeply, people usually can reach a VC of 4.5 liters (4.8 quarts) or slightly more.

The volume of air that you can force out beyond the TV is your *expiratory reserve volume* (ERV), how much air you can breathe out forcefully after you already breathed out normally. The ERV averages about 1 liter.

Subtracting the TV and the ERV from your VC gives you the *inspiratory reserve volume* (IRV). This is how much space is available in your lungs above and beyond a normal inhalation. The IRV averages about 3 liters (3.2 quarts), but you can't actually breathe that way for any length of time.

Even after you force out all the air you can, you still have a little more left in your lungs. Because they don't deflate, you have *dead space*. This is the amount of air in your lungs that is not available to participate in gas exchange. This contributes to the *residual volume* (RV), the amount of air that remains in your lungs at all times, which is 1.2 liters (1.3 quarts) on average. Add the RV to your VC and you'll have your *total lung capacity* (TLC).

## Holding your breath

You can stop breathing (hold your breath), at least for awhile, when unpleasant odors, noxious chemicals, or particulate matter is in the air around you, while you swim underwater, or just for the fun of it. The cerebral cortex sends signals to the rib muscles and the diaphragm that override the respiratory center signals — temporarily.

Holding your breath long enough to cause damage to your own brain from a lack of oxygen isn't possible. So when a little kid says he's going to hold his breath until you give him what he wants, you don't need to worry. Metabolism and gas exchange continue as usual while you hold your breath. Carbon dioxide concentration increases in the blood. At a certain point, long before brain damage is even a possibility, the chemoreceptors that work with the respiratory center are stimulated to the point where they override the cerebral cortex. At the extreme, you lose consciousness. (The evolutionarily older brainstem puts the whippersnapper cerebral cortex into time out.) You exhale and the system rapidly returns to normal.

### Speaking and singing

Speech requires breath control. The exhalation passes breath over the vocal cords, causing sound waves to be emitted, and the lips and tongue shape the sound waves into speech. The rate of exhalation is lower while you're speaking, controlled by the diaphragm, the intercostal muscles, and the abdominal muscles. Singing requires even more breath control than speaking.

### Controlling other systems

The relationship between the autonomic nervous system and the respiratory system appears to be two-way. For example, anxiety prompts hyperventilation, and hyperventilation produces symptoms of anxiety. Consciously controlling the rate and depth of breathing, mainly by achieving awareness and control of the diaphragm, has been demonstrated to decrease anxiety and sympathetic nervous system activation.

Controlled breathing is a feature of many religious, spiritual, and physical disciplines in all traditions. Clinical benefits have been demonstrated for meditation and controlled breathing in a wide variety of conditions of the neural, cardiovascular, and respiratory systems.

# Gas Exchange

So, we've got the air in the lungs. How do we get the oxygen into the blood? This process is called gas exchange and occurs only at the alveoli.

## The respiratory membrane

Each of the approximately 300 million alveoli is wrapped with capillaries, whose walls, like the alveoli's walls, contain simple squamous epithelium. Because each wall is only one cell layer thick, this tissue is well suited for the exchange of materials. The interface of the simple squamous epithelium of an alveolus and the simple squamous epithelium of a pulmonary capillary (along with its supporting connective tissue) is called the *respiratory membrane.* That's where gas exchange actually occurs. Check out the "Structures of the Respiratory Membrane" color plate in the center of the book for a diagrammatic look at this physiologically crucial interface.

# The trade-off

The respiratory membrane is where blood is reoxygenated in the process of pulmonary circulation (see Chapter 9). The process of gas exchange at the respiratory membrane is almost exactly the same as the capillary exchange process (also described in Chapter 9). Here, it's just reversed. There is more oxygen in the alveoli than in the blood, and because there are merely two layers of cells between them, the oxygen easily diffuses through.

**REMEMBER**

Though we think of the respiratory system as our means of obtaining oxygen, it's important to remember that it serves a waste management function as well. The carbon dioxide created as waste during cellular respiration is carried around in the bloodstream. Because there is less of it in the alveoli, it diffuses in as the oxygen is coming out. When you exhale, the air in your lungs is forced out, carrying this waste out with it.

Study Figure 10-2 carefully to gain an understanding of the events of respiratory gas exchange.

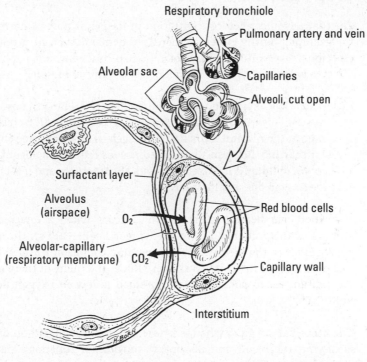

**FIGURE 10-2:**
Respiratory gases are exchanged by diffusion across alveolar and capillary walls.

*Illustration by Kathryn Born, MA*

**TIP**

The driving force of diffusion is a concentration gradient; that is, the molecules naturally move to areas where there is less of them (if they are not blocked). With gases, concentration isn't the proper measurement. Instead, it is partial pressure. In your textbook, when you see a comparison of "$pO_2$," there's no need to let that complicate the matter. In this context, you may think of it as concentration.

# Pathophysiology of the Respiratory System

The structures of the airway and respiratory membrane are in constant contact with the air, with its constant threats of temperature extremes, desiccation (drying out), harmful chemicals and particles, and pathogens.

## Hypoxemia

Fluctuations in oxygen concentration in tissues is part of normal physiology — for example, during times of increased demand such as strenuous exercise or emotional stress. In a healthy body, homeostatic mechanisms kick in rapidly and effectively when oxygen levels fall below the normal range.

*Hypoxemia* (low oxygen concentration in the arterial blood), however, is a respiratory system disorder resulting from any condition that interferes with gas exchange at the respiratory membrane, such as an obstructed (blocked) airway or alveolar scarring. Because all cells and tissues require an adequate oxygen supply to function optimally, hypoxemia harms all structures and inhibits all physiological processes to one extent or another.

A further complication of hypoxemia occurs when a low blood oxygen level stimulates the production of more red blood cells (RBCs). However, if iron is in short supply in the diet, which is a common condition worldwide, the body can't make enough hemoglobin for all these RBCs, and many of them are dysfunctional, a condition called *polycythemia.* The blood is low in oxygen despite the excess of RBCs.

The other side of gas exchange is affected, too. Blood carbon dioxide levels rise, acidifying the blood and disrupting its normal function. Many enzymes, for example, are sensitive to pH.

*Note:* Don't confuse hypoxemia with *hypoxia*. Hypoxemia can cause hypoxia, but they are technically not the same thing. Hypoxemia is low oxygen content in the blood (low supply), and hypoxia is low oxygen in the tissues.

# Airway disorders

Airway disorders are similar in their local and systemic effects. Many are chronic and all lead to hypoxemia, which may be mild, moderate, or severe.

## Asthma

*Asthma* is classified as a chronic disorder involving a "reactive airway." This means that the airway (bronchi and bronchioles) becomes inflamed (swollen) or constricted in reaction to certain triggers. The bronchial tubes may spasm. The mucosa lining the tubes may secrete excess or very thick mucus. Any or all of these conditions make breathing difficult. Treatment often involves the use of an inhaler, which delivers a *bronchodilator* to the respiratory tract, decreasing the inflammation.

Asthma is most often developed and diagnosed in childhood. Signs and symptoms of mild asthma include periods of coughing, shortness of breath, or *wheezing* (a whistling sound while breathing). The symptoms are often brought on by exercise, but sometimes they occur without that trigger. The condition may progress to moderate asthma, with more frequent and severe bouts of shortness of breath, trouble breathing while resting, and an increased respiration rate, sometimes lasting several days. In severe asthma, respiratory distress brings hypoxemia. Asthma is a chronic condition and can be fatal.

Clinically, asthma is diagnosed as *intrinsic* (triggered by factors in the body) or *extrinsic* (triggered by factors from outside the body) and treated accordingly. Intrinsic asthma may set in after a severe respiratory tract infection. Other triggers include hormonal changes, emotional stress, and fatigue. Extrinsic asthma triggers are common irritants and allergens.

## Bronchitis

*Bronchitis* is inflammation of the bronchi. *Acute bronchitis* may follow respiratory infection or exposure to irritating substances or cold temperatures. As a result, the body produces copious mucus, causing a persistent cough.

*Chronic bronchitis* is caused by long-term exposure to irritating substances, such as chemicals or cigarette smoke. The cilia of the respiratory mucosal cells are damaged and are less effective at cleaning the bronchial tissue of the respiratory mucus and the entrapped debris. Coughing develops to help the excretion of the mucus, but coughing irritates the bronchi further. The airway swells and constricts.

# Lungs

In spite of all the respiratory system's defenses, inhaled pathogens, chemicals, and particles find their way to the lung. Once there, they can be difficult to push out again.

## Pneumonia

The lung is an extremely hospitable environment for bacterial, viral, and fungal microbes. *Pneumonia* is a pathogenic infection in the lower respiratory system that results in mucus and pus building up in the airway. Pneumonia is classified by the location and extent of the infection as *bronchopneumonia*, *lobular pneumonia* (affecting part of a lobe), or *lobar pneumonia* (affecting the entire lobe).

Pathogens enter the respiratory system easily in inhaled air. A healthy respiratory system working with the immune system can eliminate almost all of them. Macrophages (see Chapter 13) patrol the alveoli to look for pathogens that make it past the mucus traps and eliminate them. As the body ages, both respiratory tissue and immune system functionality decline. Before the arrival of effective antibiotics in the pharmacological arsenal, bacterial pneumonia was a common cause of death in the elderly. (It even had a nickname, "the old man's friend," implying that a relatively quick death by pneumonia was a blessing.)

## Tuberculosis (TB)

*Pulmonary tuberculosis* is an infectious disease of the lungs caused by the bacterium *Mycobacterium tuberculosis.* The bacterium may infect the lung without symptoms for a period of several months to many years. The infection may spread to the bones and other organs. The long asymptomatic period, during which the infected person can infect others just by breathing, is what makes TB such a stubborn public health problem.

In tuberculosis, areas of bronchial and lung tissue become inflamed and die, leaving a hole in the tissue from which air can leak. Air leaking out from the lung causes the lung to collapse.

Due to its structure, *M. tuberculosis* evades efforts by the immune system to eliminate it. The immune system responds to its presence as it does for any infection: White blood cells and macrophages rush to the site of infection, and the macrophages engulf the bacterial cells and carry them to the lymph nodes (see Chapter 13). But the macrophages then behave oddly — they clump together, forming *tubercles*. Wherever the tubercles lodge, the surrounding tissue is killed, and scar tissue forms around the tubercle. Eventually, the lymph nodes become inflamed and may rupture, allowing the bacterium to spread to surrounding tissue. Because lymph nodes exist all over your body, TB can easily spread from your lungs to other areas.

## Emphysema

*Emphysema* is a *chronic obstructive pulmonary disorder* (COPD). Commonly known as a smoker's disease, emphysema can also affect people who have had long-term exposure to pulmonary irritants, such as chemicals, asbestos, or coal. Eventually, the cumulative damage to the bronchioles causes them to collapse, trapping air inside them. The pressure of the trapped air can rupture the tiny alveoli, destroying the respiratory membrane in that area. The destroyed tissue may be replaced with inelastic scar tissue (fibrosis). The elasticity of the lungs is decreased, making it hard for a person with emphysema to breathe *(dyspnea)*. As the disease progresses, the delivery of oxygen to the blood is compromised, resulting in hypoxemia.

Chapter **11**

# The Digestive System: Beginning the Breakdown

Just how does that pile of steak, potatoes, and salad on your plate become the tissues of your body? The digestive organ system gets it halfway there. (The cardiovascular system does most of the rest.)

Living systems constantly exchange energy. Physiological processes — anabolic, catabolic, and homeostatic — require energy. Ultimately, that energy comes from the light energy that plants use to transform carbon in the atmosphere (as $CO_2$) into biological matter (as carbohydrates) in the process of *photosynthesis.* Humans get their energy by consuming this biological matter, either directly or by consuming other organisms that do. The digestive system takes apart this biological matter step by step and transforms it into a form that human cells can use. In this chapter, we explain the ins and outs of the process and the organs responsible for the chore.

# Functions of the Digestive System

Digestion itself is a middle step. Other important functions of this organ system come before and after:

>> **Ingesting:** Although all animals *ingest* — take something into the body through the mouth — only humans, and possibly some of the great apes, appear to enjoy food as they ingest it.

As we discuss in Chapter 7, the perception of subtle flavors is more closely connected to olfaction than digestion. The perception of the five basic flavors is also considered *neurosensory.* Perception in the mouth is more about texture and is closely related to food's protein and fat content. This concept is captured by the food industry term *mouth feel.* These sensory perceptions guide you in the selection of foods.

Sometimes, though, ingestion isn't a feast for the senses — nothing delicious is available. The body still needs calories, though, so whatever's available is chewed and swallowed just the same. (See the section "Starting with the mighty mouth" later in the chapter.)

>> **Digesting:** Eating is fun, and ingestion is bearable, but neither provides biological molecules that your cells can use. That task is accomplished by the interaction of physical and chemical forces. The digestive tract is a muscular tube lined with chemical factories that operate under the direction of their own dedicated neural structures and under hormonal control (see the upcoming section "Structures of the Digestive System").

The digestive system processes everything down the same track, extracting fuel, biological molecules, and micronutrients from whatever you eat. (See the section "Moving through the intestines" later in the chapter.)

>> **Exporting nutrients to the body:** The end products of digestion are biological molecules such as glucose that are absorbed across the digestive membrane into the blood and then distributed in the body (see Chapter 9 for an overview of the cardiovascular system).

>> **Eliminating:** The elimination of digestive waste is part of digestion. Other organ systems have evolved to make use of the digestive system's structures to eliminate metabolic wastes of other kinds. As they say, one big pile is better than two little piles.

# The Alimentary Canal

The *digestive tract*, also called the *alimentary canal* ("alimentary" means food), or the *gastrointestinal (GI) tract*, is a tube through which ingested substances are pushed along for physical and chemical processing. The tube walls are made up of an outer fibrous layer, a muscular layer, a supportive connective tissue layer, and an inner layer (containing an epithelial lining), called the *digestive mucosa*. All the layers vary in thickness from one place to another along the digestive tract. The space inside the tube is the *lumen*, and its size varies, too.

The digestive system's gross anatomy (no pun here) is comparable to that of an industrial smelter. Some structures bring in raw materials; other structures extract, process, and ship out specific substances; and still other structures export the unused part of the raw materials back into the environment. The body uses both mechanical and chemical mechanisms to break down the raw materials and export products to the larger system (the economy in the case of the smelter; the organism in the case of the intestine). These efficient systems are organized linearly — things keep moving along in one direction at a steady pace.

**TECHNICAL STUFF**

Strictly speaking, the lumen isn't "inside" the body. Rather, the body itself is wrapped around a small piece of the environment — that is, the lumen. Neither the food that enters your mouth nor any of the partially digested substances that your digestive tract produces are inside the body, either. Fully digested biological molecules extracted from these substances and transformed into molecules that are usable by human cells cross out of the lumen and into the blood. At that point, they're inside the body.

As you read about the organs that break down food to nourish your body, refer to the "Digestive System" color plate in the middle of the book to see where the organs are located.

## Examining the walls of the digestive tract

The upper third of the esophagus contains skeletal muscle. Beginning in the middle third of the esophagus and extending to the *anal sphincter*, layers of smooth muscle comprise the digestive tract. This smooth muscle contracts in pulsating waves, pushing the lumen's contents along in a single direction. This constant wave-like contraction is called *peristalsis*.

A mucous membrane lines the digestive system, running continuously from the mouth all the way through to the rectum. This membrane protects your digestive organs from the strong acids and powerful enzymes secreted in the digestive system. The membrane's innermost cells (next to the lumen) are among those cells that are continuously replaced (see Chapter 2 for more on cell renewal).

The digestive mucosa secretes mucus to keep everything in the digestive tract moist, soft, and slippery, protecting the membrane and its underlying structures from abrasion and corrosion. The mucosa contains tissues and cells that secrete other substances as well, including gastric acid, hormones, neurotransmitters, and enzymes. The digestive mucosa also contains an extensive network of lymphatic tissue, as we discuss in Chapter 13.

The digestive mucosa takes an active role in the final stage of digestion. It delivers the products of digestion from the small and large intestine to the blood for distribution through the body. Every molecule that enters the bloodstream passes through the digestive mucosa.

**TECHNICAL STUFF**

The mucosa is continuous throughout the organs of the alimentary canal, but its structure varies. One important difference is in surface area. In the esophagus, the mucosa is smooth because its main role is merely transportation. In the small intestine, the mucosa is wavy, forming peaks called *villi* to increase the surface area; this provides more opportunity for absorption into the bloodstream despite the limited space.

## Starting with the mighty mouth

Your mouth is the starting point of your digestive system, the gateway to your other digestive organs. Besides making eating a fun experience, your mouth *(oral cavity)* serves some important digestive functions.

### Talking about your teeth and gums

Humans have 32 teeth — 16 on the top and 16 on the bottom. Your teeth begin mechanical digestion by tearing and grinding food into pieces small enough to swallow. They come in four basic types: incisors for biting, canines for tearing (especially meat), and premolars and molars for grinding.

The *gingiva* (gums) hold teeth in position, and a binding material called *cementum* embeds your teeth's roots in your jawbone (see Figure 11-1). Blood vessels that run through the jawbone and up into the pulp of the tooth supply the teeth with blood. *Dentin,* a bonelike material, covers the pulp, and an extremely hard protective enamel covers the dentin.

**TECHNICAL STUFF**

Many jawbones have difficulty accommodating the last molar in each row of teeth. Typically, these *wisdom teeth* erupt far later than the other permanent molars, usually in mid to late adolescence, and often cause jaw pain and dislocation of other teeth. Quite often, they don't erupt at all and remain "impacted" in the jawbone, which can also cause jaw pain.

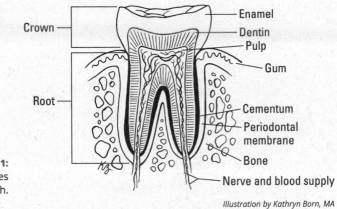

**FIGURE 11-1:**
The structures
of a tooth.

Crown
Root

Enamel
Dentin
Pulp
Gum
Cementum
Periodontal
membrane
Bone
Nerve and blood supply

*Illustration by Kathryn Born, MA*

## Your tongue helps out

Your tongue is mainly skeletal muscle tissue. The muscle is covered on the upper surface by a mucous membrane, in which are embedded taste buds. (See Chapter 7 for more on taste buds.) The tongue muscles move the food around in your mouth to assist chewing. The mucus moistens and lubricates the *bolus*, the technical term for a mouthful of food in the process of being chewed.

Muscles attach your tongue to your skull bones, and a mucous membrane on the tongue's underside attaches your tongue to the oral cavity floor. That stringy piece of membrane that you see when you touch your tongue's tip to the roof of your mouth is the *lingual frenulum*.

## The buccal membrane

The *buccal membrane* is that portion of the digestive mucosa that lines the inside of the mouth. Several *salivary glands* have ducts that course through the buccal membrane and secrete mucus and *salivary amylase,* a digestive enzyme, into the oral cavity. These glands often go into action before you take the first bite of your meal. A delicious aroma or even just the anticipation of eating something you enjoy can get those juices flowing.

**TECHNICAL
STUFF**

The enzyme salivary amylase turns starch into sugar as you chew, beginning chemical digestion.

## Pharynx and esophagus: Not Egyptian landmarks

The *pharynx,* better known as your throat, leads to the *esophagus,* the tube that extends from the mouth to the stomach. When you swallow, the bolus bounces off a piece of cartilage called the *epiglottis* preventing entry into the *trachea*, routing it into the esophagus (see Figure 11-2).

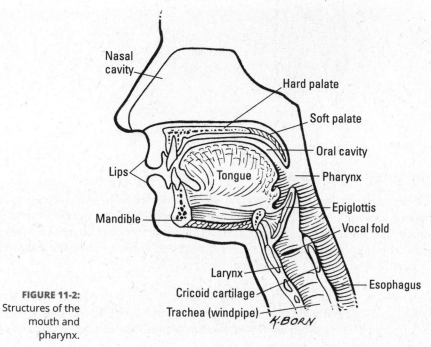

*Illustration by Kathryn Born, MA*

The esophagus has two sphincters — one at the top and one at the bottom — that control the movement of the bolus into and out of the esophagus. The *upper esophageal sphincter*, composed of skeletal muscle, is usually contracted, keeping air out of the digestive tract. It opens to allow the bolus into the esophagus and peristalsis propels it along. The lower esophageal sphincter surrounds the esophagus just as it enters the stomach. It is also closed most of the time, preventing any stomach content from moving backward.

## Stirring it up in your stomach

The outside of the stomach is a tough connective tissue layer called the *serosa*. Beneath the serosa is the *muscular layer*, which has three layers of smooth muscle fibers — oblique, circular, and longitudinal — that contract in different directions. Stretch receptors in this layer send nerve impulses to the brain when the stomach is full. These two layers support the structure of the stomach as a hollow organ.

When the lower esophageal sphincter relaxes, the bolus drops into the stomach, the widest and most flexible part of the alimentary canal. Food remains in the stomach about two to six hours, during which it's churned in an acidic substance that the stomach secretes called *gastric juice,* ground up by thousands of strong muscle contractions. The muscular walls of the stomach contract rhythmically, similar to peristalsis, but propelling the food back and forth. The stomach is the only part of the alimentary canal that has a third muscle layer (oblique) to allow for the mixing movement. The "Stomach" color plate in the center of the book shows the various parts of the stomach.

The stomach lining has two mucosal layers, the *submucosa* and the *mucosa.* The submucosa contains nerves and blood vessels to provide control and resources to both the muscle and mucosal layers (this pattern is seen throughout the alimentary canal). Gastric glands in the mucosa secrete enzymes to break down large molecules, preparing them for absorption (see Table 11-1), as well as the other components of gastric juice. The mucosa is corrugated (like cardboard), increasing the surface area inside the hollow. The folds are called *rugae.* As the stomach fills, the rugae smooth out, allowing the stomach to expand.

**REMEMBER**

The stomach's muscular action is part of physical digestion, like chewing, swallowing, and peristalsis. Breaking the food up into smaller pieces is essential to allow access to all the nutrients; if it stayed in a big ball, we'd only be able to absorb the nutrients on the surface. But don't overlook the stomach's contribution to chemical digestion — it's what really helps break down the food you eat.

As the stomach churns the bolus in the gastric acid, the material turns into an oatmeal-like paste called *chyme.* The chyme squirts into the small intestine through the *pyloric sphincter,* between the lower part of the stomach, called the *pylorus,* and the top of the small intestine, called the *duodenum.* This prevents too much of the chyme from entering at once.

**TABLE 11-1**    ## Digestive Enzymes

| Source | Enzyme | Targeted Nutrient |
| --- | --- | --- |
| Stomach | Pepsin | Proteins |
| Stomach and small intestine | Lipase | Fats (specifically butter fats) |
| Small intestine | Peptidase | Peptides |
| Small intestine | Sucrase, maltase, lactase | Sucrose, maltose, lactose (disaccharide sugars) |

# Moving through the intestines

The *intestine* is a long muscular tube (up to about 20 feet, or 6 meters) that extends from the pyloric sphincter to the anal sphincter. How does 20 feet of tubing fit into a relatively small space that's also crowded with other organs? It becomes narrow and convoluted. The intestines are classified as small and large based on their width, not their length (like hoses). The lumen of the small intestine is about 1 inch (2.5 centimeters) in diameter; the large intestine is about 2.5 inches (6.4 centimeters).

Overall, the intestine specializes in the import and export of biological substances of many kinds. As is usual for organs with import-export functions, the intestine has structures that maximize the surface area available for the exchange.

The intestine's muscular outer walls lie coiled closely together within the abdominal cavity, held in place by the fibrous sheets of the *peritoneum*. With two layers of smooth muscle tissue, longitudinal and circular, the intestine specializes in strong, sustained peristalsis. Connective tissue between the curves of the small intestine, called *mesentery*, help prevent it from bunching up or kinking.

The intestinal mucosa is continuous with the rest of the digestive mucosa. It's studded with specialized "work areas" that produce hormones, neurotransmitters, enzymes, and other substances integral to the digestive process.

The capillary beds found in the submucosa of the intestine define the interface of the digestive and cardiovascular systems. These capillaries are arrayed more or less continuously along the intestine's lumen.

The lumen is lined by the *villus* (plural, *villi*), a structure that's specialized for import and export processes and that's characteristic of tissues in body locations where substances are exchanged. *Villi* are fingerlike projections of the mucosa that multiply the surface area available for exchange, much like wharves and piers extending into a harbor increase the area for harbor activities.

Villi line the entire length of the small intestine, projecting out into the lumen. Each villus has its own assigned capillary for absorbing materials from the intestine into the blood (flip to Chapter 9 for more on the cardiovascular system). *Microvilli* are even smaller projections on the epithelial cells of the mucosa.

Some of these processes require *active transport* — the expenditure of some energy in the form of ATP. For more on active transport and ATP, see Chapter 3.

## Investigating the small intestine

The small intestine does a lot of the physical work of the digestive system, beginning with peristalsis. The small intestine is also majorly involved in digestive chemistry. It is an endocrine gland as well as a digestive organ, producing and secreting hormones that control digestion. Cells in the small intestine's walls secrete the hormones *secretin* and *cholecystokinin* (CCK), which stimulate the release of digestive fluids such as bile from the gallbladder and pancreatic juice from the pancreas.

The small intestine is divided into three structures along its 10- to 20-foot (3- to 6-meter) length: the *duodenum* (about 1 foot long, or 0.3 meter), the *jejunum* (about 3 to 6 feet, or 1 to 2 meters), and the *ileum* (about 6 to 12 feet, or 2 to 4 meters). It is approximately 1 to 2 inches (2.5 to 5 centimeters) in diameter.

The small intestine can measure around 50 percent longer at autopsy because of the loss of smooth muscle tone after death.

The role of the duodenum is to complete chemical digestion. *Brunner's glands* in its lining secrete mucus and bicarbonate directly into the lumen to help neutralize the gastric juice in the chyme (most enzymes require a near neutral pH). Other cells secrete digestive enzymes (refer to Table 11-1) that work with the bile and pancreatic enzymes to break the large molecules down into absorbable pieces.

The pyloric sphincter controls the release of chyme into the duodenum by the *enterogastric reflex*. The rate of flow is limited by the ability of the duodenum to neutralize the strong acid. The processes of chemical digestion run furiously. The carbohydrates, proteins, and fats are broken down into molecules such as glucose, amino acids, fatty acids, and glycerol. Peristalsis moves the almost-completely-digested chyme along into the jejunum and ileum, which are specialized for absorption.

The body handles the two products of fat digestion, fatty acids and glycerol, a bit differently. Short-chain fatty acids are shuttled directly to the capillary through the villus. Long-chain fatty acids are transported through the villus to the lymphatic system. In the cells, long-chain fatty acids are assembled into compounds called *triglycerides.* Glycerol is absorbed by the liver and is either converted to glucose or used in glycolysis (breaking down glucose into energy).

By the time the chyme works its way through all three parts of your small intestine, the nutrients that your body needs have been absorbed into the blood. At the ileum, the indigestible matter passes into the large intestine.

## The work of the large intestine

Chyme oozes from the small intestine to the large intestine (also called the *colon*), passing out of the ileum through the *ileocecal sphincter* into the *cecum*, the first portion of the large intestine. The material is now called *feces.*

The large intestine is about 6 feet long (almost 2 meters) and is positioned anatomically like a "frame" around the small intestine. Beyond the cecum, the large intestine moves upward as the *ascending colon*, across as the *transverse colon*, and downward as the *descending colon* and finally into the *sigmoid colon*.

In the large intestine, water is reabsorbed from the feces by diffusion across the intestinal wall into the capillaries. Because electrolytes are dissolved in water, the large intestine functions to absorb those, too (but not nutrients). The removal of water compacts the indigestible material in the colon, and with the addition of mucus, forms the characteristic texture of the feces.

In addition to undigested food, the feces contain the remnants of digestive secretions like bile. The brown color of feces comes from the combination of greenish-yellow bile pigments, bilirubin (also from bile), and bacteria.

Your intestines are home to unimaginably large numbers of bacteria, including hundreds of species. Trillions of tiny (prokaryotic) cells ingest some of the undigested material in your feces, producing molecules that have a well-known odor. (It's nothing to be embarrassed about, and nothing to be proud of, either.) Some of these bacteria produce beneficial substances like vitamin K, which is necessary for blood clotting. These substances are absorbed through the intestinal wall and transported into the blood via the capillaries. Check out Chapter 17 for more on microbes.

### Passing through the colon and rectum

As the colon completes its work, peristalsis moves feces into the *rectum*, which is located at the bottom of the colon. Stretch receptors in the rectum signal to the brain the need to defecate (release feces) when the rectum contains about 5 to 8 ounces (142 to 227 grams). Pushed by peristalsis, the feces pass through the *anal canal* and exit the body through the anal sphincter.

# Accessory Organs

The pancreas, liver, and gallbladder are often referred to as the *accessory organs of digestion*. They're not part of the digestive tract; they never come into contact with ingested material, and they take no part in the mechanical aspects of digestion. They produce and make available to the digestive tract's organs some of the chemical and biological substances that assist in digestion's chemical aspects.

# The liver delivers

The *liver* is one of the most important organs, not just in digestion but in many other functions. For example, it detoxifies the blood, metabolizes drugs, and regulates the supply of numerous molecules in the blood. The liver's digestive function is the production and transport of *bile*, one of the digestive chemicals.

**TIP**

Many of the terms related to the liver's structures and functions contain the prefix *hepato*−, meaning "liver."

## Liver anatomy

Your liver is both the largest internal organ and the largest gland in the human body. A healthy adult human liver weighs about 3 to 3.5 pounds (1.4 to 1.6 kilograms). It's located under your diaphragm and above your stomach on the right side of your abdomen (see Figure 11-3). The liver is soft, pinkish-brown, and triangular, with four lobes of unequal size and shape: the *right lobe, left lobe, quadrate lobe*, and *caudate lobe.* The liver is covered by a connective tissue capsule that branches and extends throughout its insides, providing a scaffolding of support for the afferent blood vessels, lymphatic vessels, and bile ducts that traverse the liver.

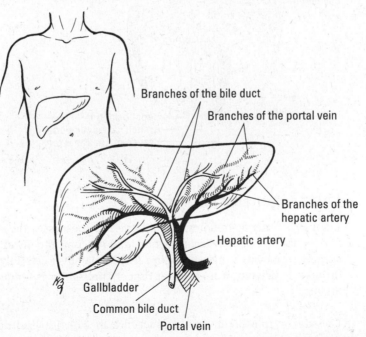

Branches of the bile duct

Branches of the portal vein

Branches of the hepatic artery

Hepatic artery

Gallbladder

Common bile duct

Portal vein

**FIGURE 11-3:**
A closer look at the liver.

*Illustration by Kathryn Born, MA*

The liver receives oxygenated blood through the *hepatic artery*, which comes from the *aorta*. It receives nutrient-rich blood through the *portal vein*, which carries blood from the capillaries of the digestive tract. This *hepatic portal system* allows the liver to process everything absorbed from the digestive system. Three hepatic veins drain deoxygenated blood from the liver, exiting the liver at the top of the right lobe and draining into the *inferior vena cava.*

Each of the four lobes is made up of tiny lobules, about 100,000 of them in all. The *hepatic lobule* is the liver's functional unit (see Figure 11-4). Each lobule is made up of millions of hepatic cells and bile canals and is supported and separated by branches of the capsule. At the lobule's vertices (peaks) are regularly distributed *portal triads* that contain a bile duct, a terminal branch of the hepatic artery, and a terminal branch of the portal vein. The *hepatocytes* are in a roughly hexagonal arrangement, with a vein in the center that carries the lobule's products out into the blood. On the surface of the lobules are ducts, veins, and arteries that carry fluids to and from them.

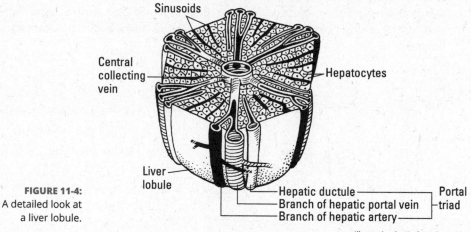

**FIGURE 11-4:**
A detailed look at a liver lobule.

*Illustration by Kathryn Born, MA*

## Liver regeneration

Among its other astounding powers, the liver possesses the ability to regenerate and regain its original size, structure, and function rapidly after partial resection surgery (removal) or massive injury. The human liver can fully regenerate from as little as 25 percent of its original tissue. This ability is unique among the major organs.

Live-donor transplantation, a procedure in which a healthy person donates a portion of his or her liver to a recipient whose liver is failing, has been performed successfully since 1989. Typically, the liver doubles in size in both the donor and recipient within only three to four weeks. Rapid replication of hepatocytes is the mechanism of growth.

## Bile production and transport

The liver produces bile, a major factor in the digestion of fats and lipids of all kinds. The bile that some of the lobules produce is collected in bile *canaliculi*, which merge to form *bile ducts*. The *intrahepatic* (within the liver) bile ducts eventually drain from the liver through the *common hepatic duct*.

Bile can be released into the duodenum from the liver, but most often it heads to the *gallbladder* for storage. Your gallbladder is a pear-shaped sac tucked into the curve of your liver whose only function is to store bile and deliver it on demand to the small intestine. Bile both enters and leaves the gallbladder through the *cystic duct*. The cystic duct joins with the common hepatic duct to form the *common bile duct*, which releases bile into the duodenum.

## Other functions of the liver

The liver functions in many other ways, affecting other organ systems. Here's a brief overview of its many functions:

>> It processes and eliminates toxins. Toxic byproducts of some drugs, including alcohol, and other substances arrive from the digestive organs through the portal vein.

>> It processes and eliminates metabolic waste. The liver removes dying red blood cells from the blood and converts the hemoglobin to *bilirubin* and other byproducts. These are used to form bile. (The iron is recycled.)

>> It stores glucose in the form of *glycogen* and reconverts it when blood glucose levels get low. This function is mediated by insulin and glucagon. (Turn to Chapter 8 for more on the endocrine system.)

>> It stores vitamins and minerals.

>> It produces many kinds of protein, including protein hormones, the plasma proteins, and the proteins of the clotting cascade (see Chapter 9) and the complement system (see Chapter 13).

# Pancreas

The *pancreas* sits in the abdominal cavity next to the duodenum and behind the stomach. We discuss the endocrine function of the pancreas in Chapter 8. The pancreas produces *pancreatic juice*, which is full of pancreatic enzymes, in response to CCK from the small intestine. Refer to Table 11-2 for information about pancreatic enzymes. Cells in the pancreas called *acinar cells* secrete pancreatic juice and passes it through the *pancreatic duct* into the duodenum. Cells along the pancreas's internal ducts secrete bicarbonate (which is alkaline) in response to secretin (from the small intestine) to neutralize the acidic chyme.

**TABLE 11-2** **Pancreatic Enzymes**

| Enzyme | Targeted Nutrient | Result of Breakdown |
|---|---|---|
| Trypsin | Proteins | Peptides (chains of amino acids) |
| Peptidase | Peptides | Individual amino acids |
| Lipase | Fats | Fatty acids and glycerol |
| Nuclease | Nucleic acids (DNA, RNA) | Nucleotides |
| Amylase | Carbohydrates | Glucose and fructose |

# The Breakdown

Each part of the digestive system has its characteristic fluid, each a complex mixture of water, electrolytes, and biological substances with a specific role in digestion.

>> **Mucus:** Every inch of the digestive tract has mucus glands whose secretions keep everything in the digestive tract moist, soft, and slippery as well as protecting the lining from abrasion and corrosion.

>> **Saliva:** Saliva (or spit) is a clear, watery solution that the salivary glands produce constantly in your mouth. You produce about 2 to 4 pints (1 to 2 liters) of spit every day. Saliva moistens food and makes it easier to swallow. It's also a component of the sense of taste — a food substance must be dissolved in the watery solution for its chemical signals to act on your taste buds (see Chapter 7).

Enzymes in saliva start starch digestion even before you swallow food. The combination of chewing food and coating it with saliva makes the tongue's job a bit easier — it can push wet, chewed food toward the pharynx (throat) more easily.

Saliva cleans the inside of your mouth and your teeth. The enzymes in saliva also help to fight off infections in the mouth.

>> **Enzymes:** Thousands of enzymes are involved in digestion. Enzymes are specialized in their function — a given enzyme typically catalyzes one or only a few specific reactions. Digestive enzymes specialize in reactions that take specific molecules apart into component chemical entities. They can be broadly classified as *proteinases* and *peptidases, lipidases,* and various kinds of *carbohydrate-active* enzymes. Enzymes are part of the digestive fluids gastric juice and pancreatic juice.

**REMEMBER**

The suffix *–ase* indicates an enzyme that breaks a molecule apart.

>> **Gastric juice:** Gastric juice is secreted from millions of tiny gastric glands in the gastric mucosa and enters the hollow of the stomach through *gastric pits* on the mucosa's inner surface. Gastric juice contains *hydrochloric acid* (HCl), which is extremely acidic and kills bacteria that may have entered the body with food. It also contains the powerful proteinase *pepsin,* which can work only in this highly acidic environment.

>> **Pancreatic juice:** Pancreatic juice contains many types of digestive enzymes. Refer to Table 11-2 for some details.

>> **Bile:** Bile, also called *gall,* is a very alkaline, bitter-tasting, dark-green to yellowish-brown fluid produced by the liver. Bile may remain in the liver or be transported to the gallbladder for storage before being expelled into the duodenum.

The physiological function of bile is to emulsify fats — that is, to create an environment in which lipid-based substances can be mixed in a watery matrix for transportation and to make them available for chemical reactions to break them down. Bile's high alkalinity helps neutralize the strongly acidic chyme that comes into the duodenum from the stomach. Another purpose of bile is to help absorb the fat-soluble vitamins K, D, and A into the blood.

**TECHNICAL STUFF**

The color of bile comes from the *bilirubins* and biliverdins that come from red blood cells (RBCs) dismantled in the liver. These pigments are deposited in the bile for elimination through the digestive tract. They play no part in chemical digestion, though recent research seems to support a role as antioxidants.

>> **Hormones:** The hormone *gastrin* regulates the secretion of HCl, mucus, and *pepsinogen* (which is turned into pepsin). As long as gastrin is flowing, your stomach continues to secrete gastric juice.

Cells in the small intestine's walls secrete the hormones cholecystokinin (CCK) and secretin, which stimulate the release of bile and pancreatic juice.

>> **Buffers:** To lower the extreme acidity (raise the pH) of gastric acid and create a more hospitable environment for most digestive enzymes, the small intestine and the pancreas secrete *sodium bicarbonate,* the same compound found in baking soda.

# Pathophysiology of the Digestive System

This section gives you some info on several common diseases and disorders of the digestive system.

## Diseases of the oral cavity

Bits and pieces of food that remain in the mouth promote the growth of the normal bacteria present in the mouth. The bacteria in the mouth make themselves right at home, secreting a gelatinous matrix called a *plaque* into the spaces between the *gingiva* (gums) and the teeth. Within about a day, the secreted material hardens into *calculus* or *tartar* (both terms mean, basically, "hard stuff"). The accumulation of plaque and the overgrowth of the bacteria cause *gingivitis* (inflammation of the gingiva). Gingivitis is a major factor in tooth loss. Many otherwise healthy people live with this chronic, low-level inflammation, a risk factor for cardiovascular and other diseases, as we discuss in Chapter 13.

Over time, the acid byproducts of the bacteria's metabolism erode the teeth's enamel, creating a "cavity." If the erosion continues, pathogenic bacteria can make their way into a tooth's pulp, a condition called a *tooth abscess*.

## Disorders of the stomach and intestines

The stomach and intestines take food in one end and push it out toward the other, with potential for trouble at every point along the way.

### Constipation

*Constipation* results when the large intestine absorbs too much water out of the feces, which makes the feces dry, hard, and a bit painful on exit. Almost everyone experiences constipation at one time or another, and some people suffer chronic constipation. Underlying causes can be dietary (generally, too little fiber or too little water), a lack of exercise, certain foods and beverages, certain drugs, or the slowing of intestinal activity that can come with aging. The most common cause of constipation is ignoring your body's signal to defecate. The feces remain in the colon too long, and too much water is absorbed into the intestinal lining, drying out the feces.

### Diarrhea

An excessive amount of water remaining in the feces causes *diarrhea*. Something is preventing a normal amount of water from being absorbed through the large intestine and back into the blood. Although a bout of diarrhea is merely an inconvenience to adults in the developed world, it's a major health concern in developing countries, particularly for children. Each year, diarrhea kills 1.5 million children around the world, and it's one of the most common causes of death of children under 5 years old, according to the World Health Organization.

One possible cause of diarrhea is pathogenic bacteria that infect the large intestine (not the beneficial bacteria that normally inhabit the large intestine) through contaminated food. The rate of peristalsis increases in an attempt to eliminate the pathogen quickly, without the water being reabsorbed.

Another cause of diarrhea is stress, working through the hormonally driven mechanisms of the *flight-or-fight response* (see Chapter 8). Among the effects of adrenaline, the most important hormone in this mechanism, is the stimulation of peristalsis. In life-threatening situations, or those perceived as life-threatening, the lower intestines may expel their contents suddenly, presumably to jettison unproductive weight and permit faster running or harder fighting.

However, chronic stress can stimulate the chronic release of adrenaline, moving feces out of the large intestine too fast, impairing the reabsorption of water, and causing chronic diarrhea.

## Appendicitis

Your *appendix* is a little sac attached to the cecum at the beginning of the large intestine. During the transfer of chyme from the small intestine to the large intestine, some material may flow into the appendix. Normally, this material makes its way out. But if it doesn't come out, depending on the material and how long it remains, the appendix can become inflamed or infected by intestinal bacteria, causing *appendicitis*. Fortunately, appendicitis is severely painful; the nature and location of the pain is diagnostic for the condition. (Figure out where your appendix is, just in case. This life-threatening condition is quite common and is always a medical emergency.)

In the worst case, or if left too long untreated, the appendix swells and bursts. The boundary between the large intestine and the peritoneum is breached, causing life-threatening peritonitis and shock.

## Gastric and duodenal ulcers

Gastric or *duodenal ulcers* are lesions in the mucosa of these tissues. Breaks in the thick protective mucus layer permit the highly acidic gastric juice to contact the lining's cells, causing pain and further tissue damage.

For many years, emotional stress was considered to be the underlying cause of ulcers. Physicians and physiologists reasoned that the parasympathetic nervous system of a person under stress sent signals to secrete more gastric juice than was needed, leading to excess acid. A large industry was built around gastroenterologists prescribing antacid drugs to ulcer patients. Then, in the late 1980s, a group of Australian physicians and researchers demonstrated that these ulcers were the result of a bacterial infection. The species *Helicobacter pyloris* (translation: screw-shaped rod in the stomach) is moderately infectious and is present in many people; but when paired with excess acid, the bacteria are able to penetrate the mucous layer and embed in the epithelium. (They move through the mucus and embed in the epithelium by twisting, like a screw. In most stomachs, they don't make it past the mucus layer.) A standardized antibiotic regimen is usually successful in eliminating the infection; the ulcers resolve as the mucosal lining is continuously replaced.

# Bowel syndromes

Bowel syndromes are of two types: *noninflammatory* and *inflammatory*. We discuss both types in the following sections.

## Irritable bowel syndrome (IBS)

In *irritable bowel syndrome* (IBS), irritation of the tissues results in a change in peristalsis. The rhythm of peristaltic contraction may either speed up or slow down. The patient may suffer diarrhea, constipation, or both. Stress reduction and a high-fiber diet usually are in the treatment plan. IBS is characterized as noninflammatory because autoimmunity and the inflammation response aren't induced.

**TECHNICAL
STUFF**

A syndrome is a collection of symptoms rather than a disease.

## Crohn's disease

*Crohn's disease* is an inflammatory bowel disease, meaning that the intestinal lining becomes inflamed. (The inflammation response is part of the syndrome.) The mucosal layer, the muscular layer, the *serosa* (the covering tissue), and even the lymph nodes and membranes that provide blood supply to the intestine can be affected. As the intestinal lining swells, ulcers, *fissures* (cracks), and *abscesses* (pus-filled pockets) can form. Crohn's can affect any part of the alimentary canal but is most common in the ileum.

In the disease's early stages, sufferers have diarrhea and pain in the lower right side of the abdomen. The inflammation can spread through layers of the wall, which narrows the hollow space inside the intestine.

Crohn's sufferers can develop nutritional deficiencies because of poor nutrient absorption. Usually, people with Crohn's lose some beneficial intestinal bacteria, including those that synthesize vitamin B12. This chronic deficiency can lead to a condition called *pernicious anemia.*

Unfortunately, the cause is still unknown. But, treatment for Crohn's includes dietary changes, rest, stress reduction, vitamin supplements, and medications to reduce inflammation and pain. Surgery sometimes is necessary to remove the affected portion(s) of the intestine.

### Ulcerative colitis

*Ulcerative colitis* (*colitis* is inflammation of the colon) is a fairly common inflammatory bowel disease in which the intestinal lining becomes inflamed. (The inflammation response is part of the syndrome.) Although the symptoms are similar to Crohn's, ulcerative colitis is limited to the colon and affects only the mucosa. Ulcers form in the lining of the large intestine, and the resulting inflammation leads to the production of a lot of mucus and pus. Abscesses can form in the lining, and the tissue surrounding them can become irritated, or damaged, or die. Ulcerative colitis can become a life-threatening condition.

Because of all this damage, the feces are often filled with blood and mucus. If the blood loss is severe enough, *anemia* (relative lack of red blood cells) can develop. As the disease progresses, the colon's lining thickens and develops scar tissue, so absorption of water and electrolytes is reduced.

The cause of ulcerative colitis is under study. A problem with T lymphocytes appears to negatively affect the epithelial lining of the colon. Sometimes an infection may start the process. Although stress doesn't cause ulcerative colitis, it can bring on an attack.

## Diseases of the accessory organs

The organs producing digestive chemicals are subject to the malfunction of the powerful chemicals they themselves produce, as well as to infectious diseases, nutritional deficiencies, and imbalances originating in other organ systems. Malfunction in any of these organs can have far-reaching effects in physiology.

### Systemic symptoms of liver disease

Because the liver has so many important functions in physiology, malfunction of this large organ can show up as signs and symptoms in other organ systems.

When the liver has failed, the bilirubin from aged RBCs isn't eliminated properly from the body. Some may be deposited in the skin, where it causes intense itching, the most commonly reported symptom of liver failure. Yellow coloration of the *sclera* (whites) of the eyes is a well-recognized sign of liver disease, from which the common term *jaundice* (yellowing) is derived. The bilirubin may enter the urine at the kidney, giving the urine a dark color diagnostic of liver problems. The feces are pale because the metabolite of bilirubin that gives feces their brown color isn't being produced and delivered to the large intestine.

Among the proteins made in the liver are most of the components of the *clotting cascade* (see Chapter 9) and those of the *inflammation cascade* (see Chapter 13). Impaired production of these proteins can result in bruising and excessive bleeding. Decreased concentrations of plasma proteins, especially *albumin,* soon brings on *edema* (fluid retention) in the abdomen, legs, and feet. A general loss of nutrients may result in chronic fatigue.

REMEMBER

The liver is capable of considerable self-regeneration. These problems may self-correct by homeostatic mechanisms when the cause of the liver malfunction is eliminated.

## Viral hepatitis

An inflammation of the liver is called *hepatitis.* Among the most common causes of hepatitis are the viral infections commonly named "hepatitis viruses A through E," in order of their scientific discovery. The study of hepatitis viruses is an area of active biomedical research, and more hepatitis viruses may be identified and named. The viruses themselves aren't related, but they all produce a similar set of symptoms. In the *prodromal stage* (immediate onset of infection, lasting about two weeks), the afflicted suffer from a sick feeling known as *generalized malaise,* as well as nervous system problems such as an altered sense of taste or smell, or light sensitivity. In the clinical stages, the liver is inflamed and enlarged, leading to abdominal pain, tenderness, indigestion, and accumulation of bilirubin.

The immune system discovered eons ago what medicine became aware of only recently: These viruses are all different. The immune system produces specific *immunoglobulins* (antibodies) against each virus. Some advanced diagnostic tools used in clinical diagnosis and public health identify the virus by identifying exactly which viral antibodies are present in the patient's blood.

Some of the viruses that infect the human liver have effectively resisted the efforts of the immune system to eliminate them. Chronic infection can bring long-term morbidity, including extensive scarring of the liver, and early death.

## You've got gall (stones)

*Gallstones* start as crystals of cholesterol or bile pigments forming in the gallbladder. Like pebbles in a jar, they reduce the gallbladder's storage capacity. Worse, they can block the common bile duct, causing bile pigments to back up into the blood, a condition called *obstructive jaundice.* A laparoscopic laser technique called *lithotripsy* can obliterate the gallstones.

## Painful pancreatitis

Inflammation of the pancreas *(pancreatitis)* may be mild or severe, acute or chronic. Pain associated with mild pancreatitis is centered around the navel and doesn't lessen with vomiting. In severe pancreatitis, the pain is an unrelenting, piercing pain in the middle of the abdomen.

Acute pancreatitis can easily go on to become a condition called *edematous pancreatitis* (fluid accumulation) or a condition called *necrotizing pancreatitis* (death of cells and tissues in the pancreas). The cause of both is the same: Digestive enzymes produced in the pancreas are blocked from exiting into the intestine. Inflammatory changes in the ducts usually underlie the blockage.

Without the enzymes, digestion may slow or be incomplete, which may harm the body's ability to maintain homeostasis. But worse, if the condition persists, the digestive enzymes are eventually turned loose on the tissues of the pancreas itself, destroying the body's ability to produce and deliver the digestive enzymes ever again. Unlike the liver, the pancreas does not regenerate.

Sometimes, a bout of pancreatitis impairs the body's ability to produce insulin, and diabetes results. (Turn to Chapter 8 for more on diabetes.)

IN THIS CHAPTER

» **Understanding what the urinary system does**

» **Breaking down the urinary system's parts**

» **Ruminating on the role of urine**

» **Seeing how your body maintains homeostasis**

» **Describing some urinary system disorders**

Chapter **12**

# The Urinary System: Cleaning Up the Act

The digestive system has its own means of preparing and disposing of the wastes from what we consume — but what about the wastes our body creates itself? This is the role of the urinary system.

In this chapter, we take you on a tour of the urinary system, reveal the makeup of urine, explain how the kidneys help your body maintain homeostasis, and look at some of the problems that affect the urinary system.

## Functions of the Urinary System

Most of the waste products of cellular metabolism and many other substances exit your body via your urinary system. Urine is the body's primary waste product, and urination (the release of urine to the environment) is the final step of metabolism.

>> **Doing the dirty work:** Metabolic wastes are toxic and would cause harm to the cell if they were allowed to accumulate inside. Cells continuously excrete their wastes into the surrounding fluid. As this extracellular fluid is absorbed into the capillaries, the waste substances enter the blood. While carbon dioxide is excreted via respiration (see Chapter 10), other wastes (byproducts of cellular metabolism) stay in the blood. The kidneys remove these toxic byproducts from the blood and put them in our urine (see the section "Putting out the trash: Kidneys" later in this chapter).

**REMEMBER**

The term *excretion* can mean the movement of material from the inside to the outside of a cell, as well as to the release of material from the inside to the outside of the body.

>> **Making and expelling urine:** *Urine* is a watery solution produced by your kidneys. It's the means through which the waste products of metabolism are removed from the blood and eliminated from the body. Urine also helps maintain the chemical equilibrium of the blood because harmful substances can be purposefully dumped there. The fluid is then transported and held in the bladder until it is released.

>> **Balancing water content:** Around half of your body weight is water, contained in intracellular (inside the cell) and extra cellular fluid (outside the cell). Your kidneys precisely regulate the release and retention of water in order to maintain blood volume and its chemical composition. In other words, they maintain the homeostasis of the blood (see the "Maintaining Homeostasis" section later in the chapter).

>> **Performing endocrine functions:** The adrenal glands sit atop the kidneys, but they're an independent organ. They produce numerous hormones, including some that influence the kidneys:

- **Regulating RBC production:** The hormone *erythropoietin,* which stimulates RBC (red blood cell) production *(erythropoiesis)* in the bone marrow, is produced in the kidney (and liver). Erythropoietin also has other physiological functions related to wound healing and recovery (flip to Chapter 9 for information on blood cell production).

- **Regulating bone growth:** *Calcitriol,* the physiologically active form of vitamin D, is synthesized in the kidneys. Calcitriol regulates, among other things, the concentration of calcium and phosphate in the blood, which promotes the healthy growth and remodeling of bone (Chapter 5 has more on bone growth).

- **Regulating blood pressure:** The kidneys also produce *renin,* a hormone involved in the renin-angiotensin system (RAS) that works to increase blood pressure. This is discussed in detail in the "Fluid Balance and Blood Pressure" section later in this chapter.

## OTHER MECHANISMS OF EXCRETION

Although most metabolic waste is excreted through the urinary system, some other organ systems have a role as well:

- Your lungs excrete carbon dioxide, a waste product of cellular respiration, in your exhaled breath.

- Your skin excretes some water, salts, and fats in the form of sweat.

- Your liver doesn't excrete wastes directly into the environment. However, it has a large role in breaking up dead cells into their chemical parts. The liver then routes the different substances for recycling, excretion through the kidneys, or packages them into bile, which is released into the small intestine.

# Structures of the Urinary System

The urinary system is quite compact. Unlike some other organ systems, you can identify the point where it begins and another point, not too far away, where it ends. See the "Urinary System" color plate in the center of the book.

Like the alimentary canal, the *urinary system* is essentially a system of tubes through which a substance passes, undergoing a series of physiological processes as it does so. The familiar tissue layers — an outer fibrous covering, a muscular layer, and a mucous layer lining the interior surface — are seen throughout the urinary system, beginning at the ureter. The mucus protects the tissues from the urine, which is slightly acidic.

## Putting out the trash: Kidneys

The urinary system begins with the *kidneys*, fist-sized paired organs, reddish-brown in color, located just below the ribs in your lower back. The kidney is shaped like an elongated oval; actually, it's shaped just like a kidney bean. (Coincidence?) The inner curve of the kidney is called the *hilum* where a number of vessels enter or exit, including the ureter, renal artery, renal vein, lymphatic vessels, and nerves. A connective tissue membrane called the *peritoneum*, as well as adipose (fat) tissue, attach your kidneys to your posterior abdominal wall. (The term *renal* comes from the Latin "ren," meaning "kidney.")

Beneath the peritoneum, a lining of collagen called the *capsule* encloses the kidney. Fibers of this layer extend outward to attach the organ to surrounding structures.

The kidneys are retroperitoneal. That is, they attach to the posterior side of the abdominal cavity's lining, on the outside of the cavity. Your back muscles on either side of your vertebral column help to protect your kidneys, as do your lowest ribs. Still, a hard blow to the back can damage a kidney fairly easily.

## Renal blood supply

Some blood from the abdominal aorta is diverted into the renal blood supply. The *renal artery* brings blood to the kidney for filtering. The renal vein drains filtered blood out of the kidney, emptying it into the inferior vena cava. The kidneys receive about 20 percent of all blood pumped by the heart each minute.

## Kidney tissues

Under the capsule, the kidney's various tissues are arranged in more or less concentric layers. The outermost layer, just beneath the capsule, is the *cortex.* Beneath the cortex is the *medulla,* a series of fan-shaped structures, called *renal pyramids,* alternating with the *renal columns.* The spaces between each pyramid, the columns, allow blood vessels through to reach the cortex. The pyramids are comprised of microscopic tubes that drip urine into saclike structures. The innermost layer contains these sacs and is called the *renal pelvis,* which channels the urine into the *ureter.* See the "Kidney and Nephron" color plate in the middle of the book to see the locations of these structures.

## Nephron

Microscopic in size (about a million of them exist in each kidney), the *nephron* is the kidney's filtration unit. Each nephron has two parts: the *renal corpuscle* and the *renal tubule.* The renal corpuscle also has two parts: the *glomerulus* (plural, *glomeruli*), a special kind of capillary bed derived from arterioles branching from the renal artery, and the *glomerular capsule* (or *Bowman's capsule*), a double-walled epithelial cup that partially encloses the glomerulus.

Leading away from the corpuscle, the nephron loops up, forming the *proximal convoluted tubule,* or PCT. The tube then straightens, pushing down into the medulla and looping up again (the *loop of Henle*). At the top, it loops again into the *distal convoluted tubule* (DCT). This tubule connects to a collecting duct that carries the urine through the renal pyramids into the renal pelvis.

The nephrons are surrounded by the *peritubular capillaries,* which perform an important role in direct secretion, selective reabsorption, and the regulation of water. See the "Selectively reabsorbing" section later in the chapter, and take a look at the "Kidney and Nephron" color plate in the middle of the book to see the nephron's individual parts.

Make sure you keep the *kidney capsule* and the *glomerular capsules* straight. The kidney has one kidney capsule on the outside and a million microscopic glomerular capsules inside.

# Holding and releasing

Fortunately, we have a storage tank for our urine over which we have voluntary control. When urine exits the kidneys, it is transported for storage until it is transported for release. This part of the urinary system, called the urinary tract, begins at the top of the ureter and ends with the urethra.

## Ureters

The *ureters* are tubes that transport the urine from a kidney to the bladder. The ureter emerges from the renal pelvis.

The walls of the ureter are similar in structure to those of the intestines: The muscular layer contracts in waves of *peristalsis* to move urine from the kidney to the bladder.

## Bladder

The *bladder* is a hollow, funnel-shaped sac into which urine flows from the kidneys through the ureters. The bladder has a capacity of about 20 ounces (about three-fifths of a liter). It lies in the pelvic cavity, just behind the pubic bones and centered in front of the rectum. In females, it's in front of the uterus.

Like other organs in the urinary and digestive systems, the bladder is made up of an outer protective membrane, several layers of muscles arranged in opposing directions, and an inner mucosal layer. The muscle layers contract to expel urine into the urethra. The mucosa is made up of a special kind of epithelial tissue called *transitional epithelium* in which the cells can change shape from cuboidal to squamous to accommodate larger volumes of urine (see Figure 12-1). Stretch receptors in the muscle layer send impulses to the brain when the bladder is becoming full. Flip to Chapter 3 for more details about epithelial tissue.

## Urethra

The *urethra* is the tube that carries urine from the bladder to an opening (orifice) of the body for elimination. The *mucosa* (epithelium) of the urethra is composed of transitional cells as it exits the bladder. Farther along are stratified columnar cells, followed by stratified squamous cells near the external urethral orifice.

Transitional relaxed

**FIGURE 12-1:**
Transitional
epithelium lines
the bladder.

Transitional stretched

*Illustration by Kathryn Born, MA*

The male and female urethrae are adapted to interact with the respective reproductive systems and therefore differ from each other in some aspects of anatomy and physiology.

In both males and females, a sphincter at the proximal end of the urethra (between the bladder and the urethra) keeps urine in the bladder. This ring of smooth muscle, called the *internal urethral sphincter*, is under the control of the autonomic nervous system. It opens to release urine into the urethra for urination.

Where the urethra passes through the pelvic floor, there's another sphincter made of skeletal muscle, called the *external urethral sphincter*, which is under voluntary control.

>> **Female urethra:** In females, the urethra is about 1.5 inches (3.8 centimeters) long. It runs along the anterior wall of the vagina and opens between the clitoris and the vaginal orifice (opening). The external sphincter is located just inside the exit point.

>> **Male urethra:** In males, the urethra is about 8 inches (20 centimeters) long, from the bladder to its opening at the tip of the penis, called the *urethral meatus.* The male urethra is divided into three named sections, based on anatomical structures:

- The *prostatic urethra* contains the internal sphincter and passes through the prostate. Openings in this region allow for sperm and prostatic fluid to enter into the urethra during orgasm.

- The *membranous urethra* contains the external sphincter. It is only about 1 inch (2.5 centimeters) in length.

- The *cavernous urethra,* or *spongy urethra,* runs the length of the penis, ending at the urethral meatus.

# The Yellow River

Urine is a bodily fluid with specific functions, just like blood and lymph. But unlike those, you see urine in the normal course of everyday events. Here's everything you always wanted to know about how it gets to be its own sweet self.

## Composition of urine

Urine is about 95 percent water, in which many wastes are dissolved. These include urea, other nitrogenous (nitrogen-containing) and various other compounds including electrolytes (ions). Its odor comes from ammonia and other substances derived from ammonia, like urea. See Table 12-1 for more about the nitrogenous components of urine.

**TABLE 12-1**     **Nitrogenous Components of Urine and Their Sources**

| Nitrogenous Component | Source |
|---|---|
| Urea | A byproduct of the breakdown of amino acids |
| Creatinine | A byproduct of the metabolism of *creatine* (present in large quantities in muscle cells, used to make ATP) |
| Ammonia | A byproduct of the breakdown of proteins |
| Uric acid | A byproduct of the breakdown of *nucleic acids* |

Urine is yellow because it contains *urobilinogen,* a compound formed from the breakdown of red blood cells (see Chapter 9). The normal color range of urine is some shade of yellow, from nearly clear to dark amber, depending mostly on the level of hydration. When you're abundantly hydrated, lots of water goes into the urine, making it more diluted and therefore paler. When you're less than optimally hydrated, the kidneys put less water into the urine. The urine becomes more concentrated and appears darker.

Ingested food, beverages, and pharmaceutical products influence the composition of urine. Urine contains *hippuric acid*, produced by the digestion of fruits and vegetables, and *ketone bodies*, produced by the digestion of fats. Some foods, beverages, and drugs impart color and odor to urine. Some physiological conditions and disorders affect the composition of urine, and urine has long been analyzed for diagnosis. For example, glucose in the urine is a sign of diabetes.

# THE COLOR YELLOW

Urinalysis offers a fairly complete assessment of the state of the body. However, some valuable clues can still come from the ancient physician's method — just looking at urine's color.

Dark yellow urine often indicates dehydration.

Very pale urine indicates "over-hydration." The body takes care of this excess hydration through homeostatic mechanisms, such as forming diluted urine. Water poisoning, in which a large quantity of water is consumed over a short period, is possible but extremely rare. The copious urine that's produced to eliminate the excess water causes electrolytes to be depleted, leading to problems in multiple organ systems.

Consumption of asparagus can turn urine greenish. Consumption of beets and blackberries can impart a reddish or pinkish color.

Excess B vitamin excretion may cause urine to be a yellow to light orange color. This could indicate a metabolic disorder or may simply be the effect of vitamin B supplements. Very high-dose B vitamin supplements may turn urine fluorescent yellow or greenish.

The metabolites of some medications, too many to list, can impart interesting colors to urine.

Bloody urine is termed *hematuria* and is potentially a sign of a bladder infection, dysfunctional glomeruli, or carcinoma (cancer).

Dark orange to brown color can be a sign of any of numerous diseases and disorders.

Black or dark-colored urine, referred to clinically as *melanuria,* may be caused by a melanoma.

Reddish, pinkish, or brown urine may be a sign of *porphyria,* a complex disorder. Most often, though, it's a sign of a diet high in beets.

# Filtering the blood

The renal corpuscle, which contains the glomerulus (capillaries) and the glomerular capsule, is the area of interface between the cardiovascular system and the kidney. That's where your body filters blood.

Blood enters the glomerulus via the *afferent arteriole.* Now in the glomerular capillaries, the plasma is forced out and caught by the capsule; this process is called *glomerular filtration.* The blood pressure in these capillaries is higher than in other capillaries, and it pushes the blood up against the glomerular wall with some force. The blood remaining in the capillaries exits the glomerulus via the *efferent arteriole* and continues on its path through the kidney.

The thin, permeable wall of the glomerulus acts as a filtration membrane. Plasma passes through into the capsule, bringing along small-molecule solutes, including wastes and toxins like urea and creatinine and useful small-molecule substances like glucose, amino acids, and electrolyte ions. The plasma captured by the capsule, now called filtrate, is then routed into the nephron (see Figure 12-2).

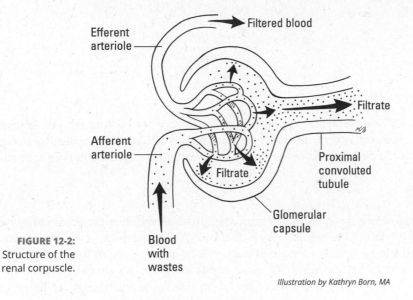

**FIGURE 12-2:**
Structure of the renal corpuscle.

*Illustration by Kathryn Born, MA*

# Selectively reabsorbing

Now, many of the wastes are in the filtrate instead of the blood. Unfortunately, so are some molecules we'd like back, such as glucose, amino acids, and electrolytes. Additionally, since about 7 liters of water are pushed out during glomerular filtration every hour, we better get most of it back because we couldn't possibly consume enough to replace it. The main role of the nephron is to get all those things back into our blood while leaving the wastes in the filtrate to be excreted as urine. As you read through this next section, refer to Figure 12-3 to help guide you.

**FIGURE 12-3:**
Nephron action.

## Microvilli

*Microvilli* line the nephrons, increasing the surface area within the tubule where substances can enter and leave the filtrate. This is also why the tubules are so convoluted. More surface area means more opportunities to get substances in or out.

## PCT

On its journey through the nephron, the filtrate first enters the proximal convoluted tubule (PCT). Here, the main goal is *tubular reabsorption* — that is, we want to pull out water and other useful molecules from the tubule so they can be placed back into the bloodstream. There are pumps along the PCT that actively transport water and ions into the cortex of the kidney. Here, the peritubular capillaries can reabsorb them.

## Loop of Henle

The filtrate then continues into the loop of Henle (nephron loop). The main goal here is water retention. Positive ions, such as sodium are pumped out and chloride ions follow. The two ions bond to form a salt, making the fluid surrounding the nephron into a hypertonic solution (see Chapter 2). As a result, water leaves the loop of Henle via osmosis and is, again, reabsorbed by the peritubular capillaries.

Water cannot be directly moved from one place to another. Our bodies do not have a means to target it. However, water readily performs osmosis — moving to where there is lower concentration (less water). We can pump specific ions into an area, and ions can form salts, so we can encourage water to follow. Increasing the solute concentration of the interstitial fluid of the medulla decreases water concentration in that area, thus water flows out of the nephron.

## DCT

Now, the filtrate enters the distal convoluted tubule (DCT). The reabsorption of water can occur here as well, but the main goal is *tubular secretion*. Some of the metabolic wastes, hydrogen ions, are too numerous to be filtered out at the glomerulus. Others, such as proteins like histamine, are too large. These wastes are still in our blood even after glomerular filtration. As a result, pumps along the peritubular capillaries target these wastes — secreting them from blood flow and into the DCT.

## Collecting duct

Finally, the filtrate enters the collecting duct. This is our last chance to get water out of the filtrate and back into our blood. Whatever is left, now urine, drips down the duct (through the medulla) and into the renal pelvis.

The amount of water reabsorbed from the filtrate back into the blood depends on the hydration situation in the body. However, even in cases of extreme dehydration, the kidneys produce around 16 ounces (about half a liter) of urine per day just to excrete toxic waste substances. Numerous other substances are returned to the blood at specific points, conserving them and maintaining the blood's chemical environment. Over 99 percent of the filtrate produced each day can be reabsorbed.

The terms "tubular reabsorption" and tubular secretion" can easily be misinterpreted. It's not what the tubes are doing. Remember the perspective — we care about the blood. We reabsorb from the tubules and secrete into them.

## Expelling urine

The urine that drips down the nephron's collecting ducts enters into the renal pelvis. From there, it moves along the ureter and into the bladder. As the urine accumulates, the pressure receptors in the lining send signals to the brain. The first message is sent when the bladder is about half full (around 6 to 8 ounces, or 177 to 237 milliliters). At a volume of 12 ounces (354 milliliters), the messages become stronger, and it becomes difficult to control the external urethral sphincter.

When the time is right to empty the bladder, the brain sends an impulse through the autonomic nervous system to open the internal urethral sphincter and contract the bladder muscles. Urine flows out of the bladder, through the urethra, and out of the body. This process is known as *micturition.*

# Maintaining Homeostasis

The chemistry of life, stunningly complex and precise, requires a tightly controlled environment to proceed optimally. The kidneys, under hormonal control, are key organs for maintaining the homeostasis of blood and other body fluids. The endocrine system has a huge range of subtle and interacting mechanisms for controlling kidney function. (See Chapter 8 for more on the endocrine system.)

## Fluid balance and blood pressure

Aspects of fluid balance that involve the blood include blood volume and the amount and nature of the electrolytes in the plasma. These factors are linked, as we discuss in the following sections.

### Blood volume (water content)

Blood volume is an important factor in circulation. The greater the blood volume, the harder the heart must work to pump the blood and the greater the pressure in the arteries. The less the blood volume, the lower the blood pressure, decreasing the amount of resources that reach the tissues. A low blood pressure will also weaken the pressure filtration in the glomeruli. However, the urinary system, when functioning properly, maintains blood volume. It does so by actually controlling electrolyte concentrations to influence water movement either into or out of the blood, even if that requires putting some strain on other systems.

**TECHNICAL STUFF**

The adult body contains about 23 pints (11 liters) of extracellular fluid, constituting about 16 percent of body weight, and about 6 pints (almost 3 liters) of plasma, constituting about 4 percent of body weight. Blood plasma and extracellular fluid are very similar chemically and, in conjunction with intracellular fluid (fluid inside the cells), help control the movement of water and electrolytes throughout the body. Some of the important ions found in interstitial fluid (fluid outside of the cells but within the tissues) are: Na$^+$, K$^+$, Cl$^-$, and Ca$^{2+}$.

## Hormonal mechanisms to control blood volume

The *renin-angiotensin system* (RAS) (also called the *renin-angiotensin-aldosterone system* [RAAS]) is called into action when stretch receptors in the kidney sense low blood pressure — especially in the glomerulus. Specialized kidney cells secrete the enzyme *renin*, which sets off a series of reactions in different organ systems, eventually resulting in the production of the hormone *angiotensin II*, a powerful vasoconstrictor. When the diameter of the blood vessels decreases (vasoconstriction), blood pressure increases. This keeps the blood circulating, giving the kidneys time to solve the volume problem.

Angiotensin II also triggers the secretion of the hormone aldosterone from the adrenal cortex, which causes Na+ to be reabsorbed from the nephrons into the blood. Where salt goes, water follows. The movement of water into the blood increases the blood volume. This, along with the vasoconstriction, raises blood pressure and thus restores the effectiveness of the glomerular filtration.

**TECHNICAL STUFF**

Some medications for *hypertension* (high blood pressure) act by blocking the production of the enzyme that leads to the production of angiotensin II. These are the so-called *angiotensin-conversion-enzyme inhibitors* (or *ACE inhibitors*). Without angiotensin II, no vasoconstriction takes place, and the kidneys don't receive the signal to increase the water content of the blood.

Furthermore, the pituitary gland is stimulated by angiotensin II to secrete ADH (anti-diuretic hormone). Sodium is reabsorbed by the nephrons but not enough water follows, so ADH causes more water to be reabsorbed from the collecting duct, which decreases the amount of urine and helps maintain normal blood volume and pressure.

*Atrial natriuretic peptide* (ANP) *hormone* is a peptide hormone secreted by heart muscle cells in response to signals from sensory cells in the atria that blood volume is too high. ANP prevents the kidneys from secreting renin. In fact, the overall effect of ANP is to counter the effects of the RAS. It acts to reduce the sodium reabsorption in the kidney. This leads to more water being released in the urine, thereby reducing blood volume and, thus, pressure.

# Regulating blood pH

The homeostatic range for blood pH is narrow; the optimum value is around 7.4. *Alkalosis* (increase in alkalinity) is life-threatening at 7.8. *Acidosis* (increased acidity) is life-threatening at 7.0. To maintain the blood pH within its homeostatic range (7.3 to 7.4), the kidney can produce urine with a pH as low as 4.5 or as high as 8.5.

**REMEMBER**

Neutral pH is 7.0, the value for water.

While the body has buffers circulating in the blood to resist pH changes, acids and, to a lesser extent, *alkalis* (bases) are byproducts of metabolic processes. The digestion of fats produces fatty acids. The carbon dioxide produced by cellular respiration can form carbonic acid if it reacts with water. Muscle activity produces lactic acid. Some acids are ingested in food and drink. The kidneys respond to changes in blood pH by excreting acidic ($H^+$) or basic ($OH^-$) ions into the urine.

Your body has buffering processes in which the kidneys are involved. A *buffer* is a type of chemical that binds with either acid or base as needed to increase or decrease the solution pH. Buffers are made in cells and available in the blood.

Three mechanisms work together to maintain tight control over blood pH:

>> Minor fluctuations in pH are evened out by the mild buffering effects of substances always present in the blood, such as the plasma proteins.

>> When sensors in the kidneys detect that the blood is too acidic, they induce the breakdown of the amino acid glutamine, releasing ammonia, a basic substance, into the blood. When the ammonia arrives at the kidney, it's exchanged for Na+ in the urine and eliminated.

>> By far the most important buffer for maintaining acid-base balance in the blood is the carbonic-acid-bicarbonate buffer. The body maintains the buffer by eliminating either the acid (carbonic acid) or the base (bicarbonate ions). Carbonic acid concentration can be reduced within seconds through increasing respiration — the excretion of carbon dioxide through the lungs increases pH. Bicarbonate ions must be eliminated through the kidneys, a process that takes hours.

**TECHNICAL STUFF**

The *pH scale* measures the concentration of hydrogen ions (H+) in a solution. The scale ranges from 0 (extreme acidity, high concentration of hydrogen ions) to 14 (extreme alkalinity, low concentration of H+, and consequently, high concentration of OH−). Solutions with balanced amounts of these two components have a pH in the neutral range.

# Pathophysiology of the Urinary System

When things go wrong in the urinary system, the healthy functioning of the entire body is affected.

## Kidney pathologies

The kidney is a complex and delicate organ, crucial to physiology but vulnerable to injury. Because of the kidneys' importance in homeostasis, kidney problems can be underlying problems in other disorders, especially those of the circulatory system.

### Calculating kidney stones

The proper name for a kidney stone is a *renal calculus,* meaning "pebble" or "hard." A kidney stone is literally a little pebble made from common waste molecules in the urine. A common cause of kidney stones is chronic dehydration. When urine is more concentrated, solutes, such as uric acid, tend to precipitate (fall out of a solution and clump together). Crystals of uric acid may develop into kidney stones, or they may travel to your joints, causing *gout* (turn to Chapter 5 for more about gout). Small stones, up to about 0.1 inch (.3 centimeters) diameter, can flow more or less quietly out through the urinary tract. Larger stones can cause considerable pain, sometimes for several days, when passing, and sometimes damage the ureters or urethra. The largest stones, 0.3 inch (.8 centimeters) or greater, frequently become *impacted* — lodged in the urinary tract and unable to be passed through the urine stream. An impacted kidney stone may cause a backup of urine in the kidney, ureter, bladder, or urethra (see the "Urinary tract obstruction" section later in the chapter) and requires medical intervention.

### Alcohol hangover

Among its many physiological effects, alcohol inhibits the release of ADH from the pituitary gland. As we discuss earlier in the chapter, ADH regulates the kidney's action in restoring water to the blood. Without ADH, too much water is excreted in the urine, and blood volume falls below the optimum range. The resulting dehydration causes dizziness, headache, and other symptoms associated with a hangover. Homeostatic mechanisms will restore fluid balance over a period of hours after the consumption of alcohol ceases. So, no matter what remedies you've heard, the only cure for a hangover is rehydrating yourself.

# Urinary tract pathologies

Obstructions, blockages, and colonization by life forms with their own agendas plague all tubal systems, biological and inanimate. At least your body can fight back against these threats to the smooth flow of urine in the urinary tract.

## Urinary tract obstruction

The urinary system's various tubes are vulnerable to an obstruction that blocks the flow of urine. The smaller the tube, the greater the risk of obstruction. The obstructing object is sometimes a stone (see the "Calculating kidney stones" section earlier in the chapter). Impacted kidney stones require immediate medical attention to avoid the risk of infection or more severe kidney damage.

## Urinary tract infection

The urinary tract is subject to infection, known as UTI. Numerous pathogenic bacteria are adept at entering the body through the external opening of the urinary tract (urethra) and establishing infection in any structure of the urinary system. A bladder infection is known as *cystitis*, a kidney infection is *pyelonephritis*, and a urethra infection is *urethritis*.

UTIs are very noticeable and uncomfortable, so patients often seek treatment immediately, before the infection spreads higher up into the urinary system. The kidney is particularly vulnerable to damage from invading bacteria. If the pathogen is able to invade the bloodstream through the close connection with the kidney, the infection, then called *septicemia*, can be extremely serious. Septicemia can lead to *septic shock*, which can be life-threatening.

## Urethral blockage

As men age, the prostate gland enlarges, usually starting around age 50. If the prostate gland enlarges too much, it can press on the urethra and block the flow of urine. The blockage can lead to UTIs in men, and it often causes painful urination (technically called *dysuria*). As the bladder becomes stressed from the blocked urethra, urine may leak out, or, more commonly, the urge to urinate becomes more frequent and sudden.

## Incontinence across the continents

*Incontinence* is the inability to control the release of urine. There are four types of incontinence:

>> **Stress incontinence:** Urine leaks out when pressure is put on the bladder, such as when a person runs, coughs, lifts something heavy, or laughs. This type is relatively common among women who have given birth (vaginally) because the muscle supporting the bladder weakens and the sphincter muscle of the urethra stretches.

>> **Urge incontinence:** Also called *overactive bladder,* this type of incontinence is a sudden involuntary contraction of the bladder's muscular wall that brings on an urgent need to urinate, possibly accompanied by a sudden release of urine. Often, these contractions occur regardless of the amount of urine in the bladder. This condition affects about 1 in 11 adults, particularly older adults. Underlying causes may be nervous system diseases, bladder tumors, bladder infection, and bladder irritation.

>> **Overflow incontinence:** In this condition, patients never feel the urge to urinate, the bladder never empties, and small amounts of urine leak continuously. This type is prevalent in older men with an enlarged prostate and is rare in women. The upper portion of the urethra passes through the prostate, so when the gland becomes enlarged, it may obstruct the passage of urine through the urethra.

>> **Total incontinence:** Some structural problems, spinal injuries, or diseases can cause a complete lack of bladder control. The sphincters of the bladder and urethra don't operate, and urine flows directly out of the bladder.

Chapter **13**

# The Lymphatic System: Living in a Microbe Jungle

Your lymphatic system is all that stands between you and a planetful of invasive microorganisms that regard you, metaphorically speaking, as a large serving of biological molecules that could feed their own processes. Your body calls on the immune processes of the lymphatic system to protect itself from invading microbes, such as bacteria and viruses, other foreign cells, and your own cells that have gone bad (such as cells that have become cancerous).

By this point, you may have wondered to yourself, "Where is the immune system?" That's a valid question, especially because many people learn that the immune system is one of the 11 body systems. However, just as the circulatory system is now called the cardiovascular system (to shift focus to the heart and blood vessels) and the excretory system no longer exists (both the urinary and digestive systems perform the function of excretion), the immune system is now redefined as a subset of the lymphatic system. Immunity is a key function of the lymphatic system, but it's not the only one. The lymphatic tissues play a vital role in circulation as well. We still have much to learn about this system. Case in point: It wasn't until 2015 that we discovered a network of lymphatic vessels in the brain!

# Functions of the Lymphatic System

The *lymphatic system* consists of a variety of components: a body-wide network of vessels and organs through which flows an important body fluid called *lymph*; a variety of very peculiar cell types; and several types of biological molecules, some of them just as peculiar. It drains and filters *interstitial fluid* (the fluid located outside of cells but within the tissues) and is the battleground for our immune defense.

>> **Confronting marauders:** Whether you're well or ill, your immune system is always alert and active. That's why none of the bacteria, fungi, parasites, and viruses that are present by the uncountable millions in the air you breathe, on surfaces you touch, and on your food, are eating you. Sure, you have lots of ways of keeping the invaders from entering your body in the first place. But disease organisms have evolved their own devious means to get past your defenses. (That's why they're "diseases" and not just "bugs.") Plenty get through. When your immune system is functioning properly, most don't stay for long. The immune system hunts them, destroys them, and wraps their cellular remains for elimination from the body.

## PATROLLING THE BORDER

Your body has several ways of keeping out invaders, both living and nonliving:

- **Skin:** This barrier keeps out innumerable invaders. Glands in the skin secrete oils that make the barrier even more effective. The *stratum basale* and *stratum spinosum* contain special cells that connect the *integumentary* system and the immune system, called *epidermal dendritic (Langerhans) cells.* For more on the skin's barrier functions, flip to Chapter 4.

- **Eyes and mouth:** The flushing action of tears and saliva helps prevent infection of the eyes and mouth.

- **Mucous membranes:** The mucosal lining of the respiratory and digestive systems traps microbes in icky sticky goo. Some of the mucosa's cells have cilia on the surface facing the lumen of the tract. These tiny, hairlike organelles move foreign matter and microbes trapped into the mucus to where they, and the mucus, are eliminated.

- **Gastric juice:** This acidic substance secreted into the stomach destroys most bacteria that are inadvertently ingested. See Chapter 11 for more on gastric juice.

- **Gut:** The beneficial bacterial species that live in the intestine can prevent the colonization of pathogenic bacteria by secreting toxic substances or by competing with pathogenic bacteria for nutrients or attachment to tissue surfaces.

>> **Stopping renegades:** The second major function of the immune system is to recognize and destroy cells of your own body that have "gone rogue" and have become potential seeds for cancerous growth. Cells go rogue every day. Most cancers can take hold only when the immune system malfunctions and fails to eliminate them, as happens with aging.

# Loving Your Lymphatic System

The lymphatic system plays a crucial role in circulation by draining the fluids that pour out into the extracellular space during capillary exchange and returning them to the blood. The lymphatic system is more than a drainage network, though. It removes toxins, helps transport fats, and stabilizes blood volume despite environmental stresses. Possibly its most interesting functions are those related to its role in immunity, fighting biological invaders. To understand the extent of the lymphatic system, turn to the "Lymphatic System" color plate in the center of the book.

## Lymphing along

In Chapter 9, you read about the cardiovascular system as a transportation network that brings substances to cells and takes their waste products away. In Chapters 8, 10, and 11, you find out how organ systems "make use of" the blood to distribute their anabolites to the target tissue. During capillary exchange, much of this cargo is pushed out through the walls of the capillaries: oxygen, ions, glucose and other nutrients, proteins, hormones, and so on, all in a watery solution. This section is about the watery solution.

The watery solution, called *interstitial* (extracellular) fluid, is the fluid in between cells, about 2 to 4 pints (1 to 2 liters) total volume at any given moment. (Interstitial means "in-between spaces.") It is essentially the same fluid you met in Chapter 9, introduced as plasma, the fluid matrix of blood.

Like plasma, interstitial fluid is continuously flowing: The pressure of the heartbeat pushes this watery solution across the capillary walls and out into the interstitial space, a total of about 50 pints (24 liters) per day. Much of it is reabsorbed into the blood at the venous end of the capillaries. The rest goes on a detour through the lymphatic system. The fluid coursing through the lymphatic system is called *lymph*. After passing through the lymphatic system, the fluid rejoins circulation at the subclavian veins, and it is plasma again.

Plasma, interstitial fluid, and lymph are all the same watery solution of plasma proteins, electrolytes (ions), and various dissolved and undissolved substances. The fluid circulates around the body. The name it's called by depends on where it is.

We need the fluid in our tissues, but we also don't want it to remain stagnant. Plus, the volume that returns to the bloodstream at the capillaries is less than what was pushed out. If we didn't drain it, excess fluid would gather, called *edema*. So, we continuously drain it, passing it through several filters before adding it back to the blood.

# Structures of the lymphatic system

The structures of the lymphatic system resemble those of other organs and systems that function to move fluids around. The lymphatic system has its own tubes, pipes, connectors, reservoirs, and filters. It lacks its own pumping organ but, like venous circulation, makes use of skeletal muscle action for this purpose. Figure 13-1 is a schematic showing the lymphatic structures and their relationship to blood flow.

## Lymphatic vessels

The *lymphatic vessels* are the tubes that carry lymph. They form a network very similar to the venous system. You could even think of the lymphatic system as an alternative venous system, because the lymph that the vessels transport comes out of the arterial blood and is delivered back into the venous blood. Like the venous system, the lymphatic system's vessels start small (the lymph capillaries) and get larger (the lymphatic vessels), and even larger (the lymphatic ducts). Like veins, lymphatic vessels rely on skeletal muscle action and valves to keep the fluid moving in the right direction. The structure of the lymphatic vessel wall is similar to that of the veins, but thinner. Lymph vessels are distributed through the body, more or less alongside the blood vessels.

## Lymphatic ducts

The largest of the lymphatic vessels, the *lymphatic ducts,* drain into two large veins. The right lymphatic duct, located on the right side of your neck near your right clavicle, drains lymph from the right arm and the right half of the body above the diaphragm into the *right subclavian vein.* The *thoracic duct,* also called the *left lymphatic duct,* which runs through the middle of your thorax, drains lymph from everywhere else into the *left subclavian vein.*

SYSTEMIC CIRCULATION / \ PULMONARY CIRCULATION

Lymph node

Lymphatic
capillaries

Lymphatic
duct

Subclavian
vein

Lymphatic
vessel

Pulmonary
capillaries
(blood)

Valve

Arteries

Lymph node

Systemic
capillaries
(blood)

**FIGURE 13-1:**
Lymphatic flow.

Lymphatic capillaries

*Illustration by Kathryn Born, MA*

## Lymph nodes

*Lymph nodes* are bean-shaped structures located along the lymph vessels (see Figure 13-2). Dense clusters of lymph nodes are found in the mouth, pharynx, armpit, groin, all through the digestive system, and other locations. Each lymph node is encapsulated (covered) by a fibrous connective tissue *capsule. Afferent* lymphatic vessels cross the capsule on the convex side, bringing lymph into the node. The node's *efferent vessel,* which carries the filtered lymph out of the node, emerges from the indentation on the concave side of the capsule, called the *hilum.*

On the inside, the capsule sends numerous extensions that divide the node internally into structures called *nodules.* A nodule is filled with a meshlike network of fibers to which *lymphocytes* and *macrophages* (another immune system cell type) adhere. As the lymph flows through the node, pathogens, cancerous cells, and other matter in the lymph are engulfed and destroyed by macrophages targeted by lymphocytes. The cleaned-up lymph travels toward the venous system in the efferent vessels.

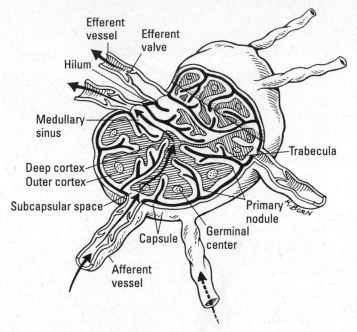

FIGURE 13-2:
The anatomy of a
lymph node.

*Illustration by Kathryn Born, MA*

The lymph nodes also provide a safe and nurturing environment for developing lymphocytes. (See the "Lymphocytes" section later in the chapter.)

TECHNICAL
STUFF

Lymph nodes are sometimes mistakenly called *lymph glands.* They don't secrete anything, so they don't meet the definition of glands.

Swelling and tenderness in the lymph nodes, especially of the pharynx, is a symptom of an infectious disease. Swollen lymph nodes ("swollen glands") is not itself a disease or pathological condition. It's the immune system doing its job.

## The splendid spleen

The *spleen* is a solid organ, located to the left of and slightly posterior to the stomach. It's roughly oval in shape, normally measuring about 1 by 3 by 5 inches (3 by 8 by 13 centimeters) and weighing about 8 ounces (23 grams). Essentially, its structure is that of a really large lymph node, and it filters blood in much the same way the lymph nodes filter lymph, removing pathogen cells along with exhausted RBCs and many kinds of foreign matter.

The spleen is enveloped by a fibrous capsule that extends inside creating chambers called *lobules.* It has a *hilum,* a spot where several different vessels cross the capsule. The spleen's hilum contains the *splenic artery, the splenic vein,* and an *efferent lymph vessel,* a similar configuration to the lymph node. Note that the spleen has no role in filtering lymph (only blood) and no afferent lymph vessels.

Inside, the spleen is divided into functional subunits by outgrowths of the capsule's fibrous tissue. Within each subunit, an arteriole is surrounded by material called *white pulp* — lymphoid tissue that contains lymphocyte production centers. Farther toward the outer edges of each compartment, similar masses called *red pulp* surround the arteriole. The red pulp is a network of channels filled with blood, where most of the filtration occurs. (It's also the major site of destruction of deteriorating RBCs and the recycling of their hemoglobin.) Both white pulp and red pulp contain *leukocytes* that remove foreign material and initiate an antibody-producing process.

### Give us a "T"

A T cell, that is. The *thymus gland* overlies the heart and straddles the trachea, sitting just posterior to the sternum. It produces *thymosin,* a hormone that stimulates the differentiation and maturation of T cells. (See the "Lymphocytes" section later in the chapter for more on T cells, and flip to Chapter 8 for more on hormones.) The thymus is relatively large in childhood; it decreases in size with age.

# Identifying Immune System Cells

Immune system cells are special in many ways. In shape and size, they're far from the compact epithelial or muscle cell types. Immune system cells have about a dozen distinctive shapes and many different sizes, and some have the ability to transform themselves into other, even weirder forms and to multiply extremely rapidly. See Table 13-1 for an overview of the different types of immune system cells.

Over the past few decades, immunologists and cell biologists have discussed immune system cells in terms of several classification systems that have not held up well on further investigation. Forming and discarding theories and classification systems based on new knowledge is expected and welcome in immunology, as in any science. The structure and physiology of immune system cells will be under study for the foreseeable future. The following discussion of specific types of immune system cells gives you some idea about some established concepts. Keep this limited goal in mind, especially when checking out Table 13-1.

**TABLE 13-1** **Cells of the Immune System**

| Cell Type | Function | Comment |
|---|---|---|
| Neutrophil | Phagocytizes bacteria. | 40%–70% of total WBCs; first responders to infection site. |
| Basophil | Defense against parasites; mediates inflammation. | 1% of total WBCs. |
| Eosinophil | Destroys antigen-antibody complexes. | 1%–4% of total WBCs. |
| Monocyte | Matures into a macrophage, which phagocytizes bacteria and viruses. | 4%–8% of total WBCs; largest of WBCs. |
| Macrophage | Phagocytizes pathogens and dead cells; stimulates the production of other WBCs. | Monocytes produce macrophages in large quantities in the early stages of the inflammation response. |
| B lymphocyte (B cell) | Produces antibodies. | Combined, B and T lymphocytes account for 20%–45% of total WBCs. |
| T lymphocyte (T cell) | Attack pathogen directly. | Ts and Bs can from memory cells for quick response if the pathogen returns. |
| NK cell | Destroys cancerous cells and cells infected by viruses. | Only lymphocyte involved in innate immunity. |
| Mast cell | Initiates inflammation response by releasing histamine at the site of injury. | Responds to allergens. |

Our numerous immune cells are constantly patrolling our bodies, floating in the bloodstream, hanging out in the interstitial fluid, moving through the lymph, and some post up in the lymph nodes acting like guards checking everyone coming through a security gate. Most often, our bodies are able to destroy the pathogen before they can unpack their bags and start making us sick. This is the strategy of the *innate immune system*. If a pathogen makes it past our first line of defense — mechanical and chemical barriers (see the "Patrolling our borders" sidebar, earlier in this chapter) — we have a barrage of cells at the ready. All but the B and T cells fall into this category. If the pathogen is especially prolific or finds a good hiding place, we start to feel the characteristic symptoms of the infection. The battle is not lost, though, we have a secret weapon: the *adaptive immune system*. We enlist the B and T cells to mount a targeted attack on the pathogen, minimizing the collateral damage to our own cells that the innate strategies inevitably cause.

## Looking at leukocytes

Immune system cells are called *leukocytes* ("white cells"), because they appear white in color under a microscope. Although all blood cells, red and white, develop from *hematopoietic stem cells* in the red marrow, the leukocytes contain no hemoglobin and no iron. Unlike RBCs, all leukocytes retain their nuclei, organelles, and cytoplasm through their life cycle. Fewer leukocytes are produced than RBCs, by a factor of around 700.

Also called *white blood cells* (WBCs), leukocytes are present everywhere and functioning at all times. You notice their presence in the acute phase of certain diseases — the immune response, not the invader directly, produces the well-known symptoms of flu. They function not only in the blood (really, in the plasma), but also in the interstitial fluid and the lymph. They're never far from a site of injury or infection because they're everywhere. When a splinter pierces your finger, a contingent of local WBCs arrives at the site instantaneously.

Sometimes it's difficult to remember that these bizarre warrior cells with their amazing superpowers are your cells — are *you* — just like your skin cells and your blood cells. Discussing them is difficult without resorting to language that makes them seem like a quasi-military force from outside your body. These cells are acutely aware (metaphorically speaking) that "they" are "self" (you). In fact, that's the primary distinction that matters to them: *self* or *nonself*. The overarching mission of a leukocyte is to protect self from other *biotic* (living) nonself, destroying the invaders when possible. They are necessary to establish more or less mutualistic relationships with other life forms as well (such as the beneficial bacteria living in our gut). It's hard not to picture a disciplined army, but that's a useful metaphor to remember when exploring the roles of the different leukocytes.

# Lymphocytes

The *lymphocytes* are one group of leukocytes (WBCs) that includes the B cells, T cells, and NK cells. These cells work in conjunction with each other during infection — each using a different mode of attack. NK cells are the only ones in this category that are an innate strategy, attacking cancerous cells and cells infected by viruses. Once activated, B cells produce antibodies and some T cells directly attack pathogens. This is our adaptive strategy and is discussed in detail in the "Adaptive Immunity" section later in this chapter. The surfaces of lymphocytes are covered with receptors, which are molecules that fit with a specific *antigen*. (Check out the "Examining Immune System Molecules" section later in the chapter for an explanation of antibodies and antigens.)

All lymphocytes originate in the red marrow from the same type of hematopoietic stem cell. B lymphocytes and NK cells leave the marrow fully differentiated and enter the blood and lymph. T cells travel to the thymus gland to complete their differentiation in an environment rich in the hormone *thymosin.* Then they move to a lymph node, where they further differentiate into one of several different cell types, each with its own function in the immune response: helper T cells, cytotoxic (cell killing) T cells, or suppressor T cells.

## Phagocytizing leukocytes

Several different types of WBCs utilize *phagocytosis* as part of their strategy, engulfing and digesting any foreign material. This strategy is discussed in further detail in the "Immune System Mechanisms" section.

### Neutrophils

*Neutrophils* are the most numerous of the WBCs (40 percent to 70 percent of the total number) and are continuously present and active in the blood and lymph. Neutrophils squeeze through the capillary walls and into infected tissue, where they phagocytize the invading bacteria. They also utilize a chemical attack by a mechanism called degranulation, which we discuss in the "Immune System Mechanisms" section.

Neutrophils are the leading candidate for the Cell with the Shortest Life Span Award. They circulate for about a day and then undergo *apoptosis* (programmed cell death). If they're signaled to an infection site, they function for another day or two, but that's it.

### Monocyte and macrophage

*Monocytes* aren't really stem cells, but they have some functions in common with stem cells — they exist to produce other specialized cells on demand. Monocytes divide to produce two other kinds of immune cells, *macrophages* and *dendritic cells.* In a homeostatic state, the monocytes replenish these cells as necessary. In response to inflammation-response-related stimuli, monocytes travel to the site and begin to turn out vast numbers of its daughter cells.

Macrophages (literally, "big eaters") are large phagocytic cells that target pathogens and dead self-cells. In the early stages of the immune response, macrophages initiate the mass production of other types of WBCs. Dendritic cells, along with macrophages, serve as the bridge between innate and adaptive immunity.

# Examining Immune System Molecules

As noted earlier in the chapter, leukocytes are difficult to classify either by structure or by function because the structure and physiology of leukocytes are stupefyingly complex and astoundingly flexible. Most leukocytes, even those that phagocytize, also produce chemicals of many kinds that modulate the functional activities of many other cell types. Immune response involves a lot of cell-to-cell communication, and leukocytes make and use proteins, enzymes, hormones, and neurotransmitters. Some of these molecules are familiar from the physiology of other organ systems. Others are special creations of the immune system and have no other functions.

# Histamine

*Histamine* is a nitrogen compound with several physiological functions but is best known for its role in local immune responses. Histamine is produced by basophils and by mast cells found in nearby connective tissues as part of the inflammatory response. It plays a major role in many allergic reactions because they get over-stimulated. As part of the inflammation response, histamine dilates small blood vessels, activates the vascular endothelium, and increases blood vessel permeability to WBCs and inflammatory proteins. It also irritates nerve endings, leading to itching or pain.

The itchy bump on your skin after a mosquito bite is caused not by the bite but by the histamine that's released to initiate the process by which the antigens that the mosquito introduced are destroyed. It's also why scratching them makes the itch worse — you damage the surrounding tissue, leading to more inflammation!

# Chemical defense

Numerous chemicals are produced by leukocytes to destroy a pathogen or to communicate with other immune cells. Table 13-2 contains a brief overview of some of these chemicals.

**TABLE 13-2**     ## Chemical Defenses of the Immune System

| Cell Type | Function |
| --- | --- |
| Collectins | Gather pathogens together for more efficient phagocytosis |
| Cytokines | A group of chemicals, including interleukins, that communicate with the other immune cells |
| Defensins | Punch holes in the cell wall/membrane of pathogens; especially useful on bacteria |
| Interferons | Block replication of viruses and cancerous cells; stimulate other immune cells |
| Perforins | Poke tiny holes in cell membranes |

# Antigens

*Antigens* are molecules that stick out from the surface of a cell. They can be proteins, carbohydrates, or a combination of the two and come in a large variety of shapes and sizes. The term *antigen* is often confused with *pathogen* (for example,

"WBCs attack the antigen"). Antigens, in fact, aren't a bad thing at all — they're actually essential. It's like every living cell is always wearing its own, personalized hat, and that's how we recognize them from non-self.

# Antibodies

An *antibody* is a type of protein molecule with immune function that is made exclusively by B cells.

Antibodies, also called *immunoglobulins* (abbreviated *Ig*), are produced in activated B cells in response to the presence of a pathogen. Only a B cell that has a receptor that matches the pathogen's antigen will be activated. It will then manufacture numerous antibodies that will only latch on to the specific pathogen for attack. The healthy human body has thousands and thousands of antibodies, each specific to one antigen, and is capable of producing large quantities of some of them on demand.

The membranes of lymphocytes are covered with receptors for thousands of different antigens, including antigens they haven't encountered. When a B cell encounters a new antigen (a membrane receptor binds it), the B cell multiplies prodigiously and (almost) all the new cells are devoted to the production and release of antibodies specific to that antigen. The antibodies circulate, efficiently binding and disabling their target. The antigen-antibody complex calls in the phagocytes and may activate the complement system. To protect you from further infection, antibodies saturate your tissues.

B cells are manufactured with random receptor shapes in hopes that any pathogen we may encounter will have a matching B cell. Statisticians argue that the chances of our having a match are so astronomically high that it's safe to say there's one in our body somewhere. The problem is finding it before the pathogen does irreversible damage.

## Antibody specificity

What makes one antibody specific to one antigen? In a word, shape.

REMEMBER

Antibodies are proteins, and like those other useful proteins, the enzymes, they do their jobs by binding very tightly to a counterpart molecule of very specific configuration.

Antibody molecules are Y-shaped, with a *binding site* on each of the short arms. A binding site, in turn, has a specific and intricate shape. An antibody can bind only an antigen that has the complementary shape. A common metaphor for this is a lock-and-key mechanism: Just as a key has a specific shape and can fit only one lock, the antibody can bind only antigens that match its shape.

## Antibody function

Antibodies work in three general ways: promoting inflammation, activating the complement system, and direct attack. Inflammation, in addition to bringing more WBCs to the area for battle, helps prevent the pathogen from moving elsewhere in the body. The complement cascade works to destroy the pathogen in numerous ways described in the next section. Antibodies' direct attack has three effects:

>> **Neutralization:** Prevents the pathogen from binding to our own cells

>> **Agglutination:** Clumps numerous pathogens together for more efficient phagocytosis

>> **Precipitation:** Makes the antigens insoluble so it's easier for a phagocytic cell to find and engulf it

## IgG antibodies

Immunoglobin molecules come in five classes, named, with great imagination, IgA, IgD, IgE, IgG, and IgM. IgG antibodies are the most important class, accounting for about 80 percent of the antibodies. They're the type most involved in the secondary immune response as they circulate in the blood and the other body fluids. IgAs are found in exocrine secretion (like tears and bile), while IgDs are found on the surfaces of B cells (forming the receptors). IgMs are specialized for blood compatibility, and IgEs promote inflammation. The overproduction of IgEs causes allergic reactions.

# Complement system proteins

The *complement system* supports the activity of antibodies to clear pathogens. The system is composed of about 26 proteins. The process is similar in many ways to the clotting cascade in response to blood vessel injury (see Chapter 9). In fact, some of the very same proteins are involved.

The complement system must be activated by one of two very specific mechanisms, involving either antigen-antibody complexes, called *specific immune response,* or antigens without the presence of antibodies, called *nonspecific immune response.*

The functions of complement proteins include

>> Making bacteria more susceptible to phagocytosis

>> Directly lysing some bacteria and foreign cells

>> Producing chemotactic substances (signaling molecules)

>> Increasing vascular permeability

>> Causing smooth muscle contraction

>> Promoting mast cell degranulation

# Immune System Mechanisms

The recent and continuing development of technological tools for microbiology and molecular biology has permitted observations of the previously unimagined details of the immune system's mechanisms. Our conceptual understanding of them follows a little behind. The following sections introduce you to the subtlety and complexity of immune system mechanisms.

## Phagocytosis

*Phagocytosis* is the simplest of immune response mechanisms and probably older than multicellular life: The invader or foreign matter is just surrounded and digested. Phagocytosis is probably the body's most frequently used mechanism, too, because the most numerous of the leukocytes, the neutrophils work this way. (Refer to the "Phagocytizing leukocytes" section earlier in the chapter.)

### WHAT'S ALL THE PUS ABOUT?

*Pus,* like mucus, is yucky. But pus is proof that your body is fighting an invader and that your immune system is doing its job.

You could say that pus is a "product" of the immune system. This thick, white-to-yellow goo, sometimes shot with blood, appears at a site of injury or infection. (What would adolescence be without at least a few pus-filled pimples on the skin?) Its color and texture come from the dead phagocytes that make up most of it — WBCs that have done their job and then died with the cellular remains of thousands and thousands of invaders wrapped up inside. Pus also contains some dead tissue, some blood, and some lymph. Your body eliminates pus through the colon or through the skin's pores.

# Degranulation

Various types of WBCs have *granules* in their cytoplasm. Granules aren't chemicals but little packets of chemicals, such as histamine, cellular toxins (see Table 13-2), enzymes, and other proteins. Cell biologists have identified several different types of granules with very specific chemical contents.

A process called *degranulation* moves the granules out of the cell, releasing the chemicals into the interstitial space (via exocytosis; see Chapter 3). After they've been absorbed into the interstitial fluid, these chemicals carry out a range of specialized immune functions. Some destroy invaders directly. Some regulate immune system processes.

Granules in *eosinophils* play a crucial part in the immune response to enteric (intestinal) parasites through their release of toxic proteins (our own natural pesticides!). Eosinophil numbers increase during allergic reactions and parasitic infections. Eosinophils also perform a limited phagocytic function in the destruction and elimination of antigen–antibody complexes.

Neutrophils, discussed above as phagocytizing cells, contain granules in their cytoplasm that release many powerful substances. Because neutrophils are the most numerous of the WBCs, their granulocytic properties are very important in the immune response, especially to bacterial infection.

The degranulation of *basophils* releases histamine and *heparin* (an anticoagulant). This is a source of the histamine found at the site of inflammation and allergic reactions. Like eosinophils, basophils play a role in both parasitic infections and allergies.

*Mast cells* are present in most tissues, characteristically surrounding blood vessels and nerves, and are especially numerous near the boundaries between you and the outside world, such as in the skin and in the mucosa of the respiratory and digestive systems. Mast cells are granular cells that play a key role in the inflammatory process. When activated, a mast cell rapidly releases the contents of its granules and various hormonal mediators into the extracellular space. They're involved in allergic reactions, anaphylaxis, and autoimmunity.

# Inflammation is swell

The last time you got a thorn or a splinter in your hand or foot, you may have noticed that the site of entry became red, hot, swollen, and tender. These are signs and symptoms of an *inflammation response.* The inflammation response is a basic way the body reacts to infection, irritation, or injury, a mechanism for removing the injurious object and initiating the healing process. Inflammation has recently been recognized as a type of innate immune response.

When the splinter punctures your skin, the injured cells release mediator chemicals, particularly *histamine* and *bradykinin,* which initiate the inflammation response. Histamine also activates the complement system. The induction of the first complement protein stimulates the production of others, which stimulate still others, continuing in a rapid but controlled chain reaction until the full-blown inflammation response is raging. Immune system cells, mainly monocytes and other phagocytic cells, rush to the site to fight any microbes that were carried in on the thorn or entered through the skin opening.

**TECHNICAL STUFF**

When histamine is released, the chemical bradykinin is released along with it. Bradykinin causes the nerves to send pain messages to the brain. Thanks, bradykinin, we guess.

# Adaptive Immunity

When the mechanisms of our innate immunity do not eliminate the pathogen, our adaptive mechanisms join in the battle. The trigger for this requires an *antigen-presenting cell* (APC); usually a macrophage or dendritic cell. When an APC comes across an unrecognized pathogen (meaning it has not been affected by any sort of immune response) it engulfs and digests it. However, it preserves the antigens and displays them on its own cell membrane. It then flows around our fluids hunting for a T cell that has a matching receptor, thus activating the adaptive immune response.

## Cell-mediated immunity

The matching *helper T cell* that has just been sitting around waiting to be called into action now begins to *proliferate* (make a bunch of copies of itself). The new helper T cells release *cytokines,* which activate the matching *cytotoxic T cells.* This will then proliferate into an active cytotoxic T cell and a *memory T cell* (which won't do anything at this point). The cytotoxic T cell will then bind to the antigens on the pathogens and release *perforins* to destroy it. This process is illustrated in Figure 13-3.

The T cell process, or cell-mediated immunity, creates an army of cells that will attack the pathogen that initiated the process. Their mode of attack is much like sending soldiers to the front lines to battle the enemy with weapons. It is especially effective for viruses and cancerous cells.

① Macrophage digests pathogen and displays antigen on its surface.

Pathogen

Displayed antigen

② Cytotoxic T cell binds to displayed antigen and
Helper T cell binds to displayed antigen and proliferates.

③ Helper T cell activates cytotoxic T cell.

To B cells

Cytokines

Interleukin-2

Activated cytotoxic T cell

④ Proliferation and differentiation

Cytotoxic T cell

Memory cell

Infected cell

Antigen

Hole in cell's membrane

Cytolysis

Perforin molecules

**FIGURE 13-3:** Cell mediated immunity.

⑤ a. Cytotoxic T cell binds to infected cell.

b. Perforin creates holes in infected cell's membrane

c. Infected cell lyses.

*Illustration by Kathryn Born, MA*

# Humoral immunity

*Humoral immunity,* also called antibody-mediated immunity, is our adaptive defense that utilizes the B cells. It works in conjunction with (rather than after or instead of) the cell-mediated process. The strategy here, to continue the army metaphor, is to turn the B cells into bomb-making factories.

B cells will bind to the matching antigen of a pathogen, but until they're activated, nothing else will occur. When the helper T cell releases cytokines during the cell-mediated process, they activate this bound B cell, leading to proliferation. The new cells become either *memory B cells* (which again, do nothing right now) or *plasma B cells*. The plasma B cells then go about their task of manufacturing antibodies. Figure 13-4 illustrates the process of humoral immunity.

**FIGURE 13-4:** Humoral immunity and antibody actions.

*Illustration by Kathryn Born, MA*

## Secondary immunity

It's the upside of infection and disease, at least some infections and diseases. After your immune system has dealt with certain pathogens, it produces the condition of immunity. *Immunity* is the condition of being able to resist infection by a particular pathogen because your body has defeated it before.

Both the cell-mediated and humoral mechanisms produce *memory cells* during the primary response. These cells hang out in your body, particularly in the lymph nodes, monitoring the fluid for their matching pathogen (with its matching antigen). Should these memory cells come across it, they reactivate and attack in their corresponding ways. Because this often occurs after exposure to the pathogen but before it can generate any symptoms, you're immune to it.

## Immunization

*Immunization* is the process of inducing immunity to specific antigens by inoculation. *Inoculation* is the introduction of an antigen into the body to stimulate the production of antibodies. The antigen is introduced, often by injection but sometimes orally or intra-nasally, in a preparation called a *vaccine* that contains a sample of the pathogen either killed or in an *attenuated* (alive but weakened) form. Essentially, vaccination induces a very mild primary response (so mild you may not be aware of it) so the secondary response can be activated rapidly when you encounter the antigen again in the environment. Researchers have developed vaccines for many infectious diseases and are developing more all the time.

# Pathophysiology of the Immune System

Malfunction or failure in any part of the immune system is a threat to homeostasis and continued existence in any organism. But other animals just don't have the same problems as humans do. The peculiarities of immune-related diseases in humans are due, in part, to human culture, which permits so many to survive to experience the age-related decline in immunity and requires continuous close contact between individuals that can sustain endemic or promote epidemic microbial diseases.

## The immune system and cancer

Speaking of cancer as one disease is no longer possible, but all cancers are similar in that they represent a failure of the immune system to detect and destroy malignant cells. Often, the underlying cause of immune malfunction is simply the aging of the immune system. *Malignant cells,* by definition, divide and reproduce rapidly and in an uncontrolled way. The immune system destroys those cells as fast as it can. A young and vigorous immune system may eliminate all the malignant cells. Eventually, however, the immune system slows down and makes mistakes. A malignancy that arises may steadily gain ground and finally defeat it.

The immune-compromised patient is vulnerable to cancers of many kinds. The high incidence of an otherwise rare cancer called *Kaposi sarcoma* was among the first clues to emerge of the deadly new epidemic in the early 1980s and a strong clue that the target of the new virus was the cells of the immune system (see the "HIV and AIDS" section later in the chapter).

An organ transplant recipient must be on a regimen of immunosuppressant drugs his or her whole life. Besides other debilitating side effects, these drugs make the patient vulnerable to cancers of many kinds. Some patients suffering from autoimmune diseases are treated with immunosuppressants, too, and have the same vulnerability.

# Immune-mediated diseases

*Immune-mediated diseases* are conditions that result from abnormal activity of the immune system. *Autoimmune diseases* are disorders caused by the immune system mounting an attack on its own cells. *Allergy* is, essentially, an overreaction of the immune system to a harmless substance in the environment.

## Autoimmune diseases

Like a real-life horror movie, the awesome coordinated power that stands ready to engage any and all biological invaders in deadly combat instead attacks and destroys the body's own tissues. Autoimmune diseases are an active area of basic and clinical research, but on a fundamental level, what causes the immune system to turn on its "self" remains obscure. In most cases, a combination of factors is probably at work. For example, a viral infection may activate (or deactivate the suppression of) a genetic error. Certain autoimmune disorders affect many more women than men, so hormone activity is probably a factor.

Autoimmune disorders are many and various in terms of the pathophysiology. An autoimmune disease can be relatively benign, like *vitiligo,* in which the immune system destroys *melanocytes* (pigment-making cells), resulting in white patches of skin on different parts of the body. Other autoimmune diseases are far more serious. They can affect any part of the body, including the heart, brain, nerves, muscles, skin, eyes, joints, lungs, kidneys, glands, digestive tract, and blood vessels. Clinical experts disagree among themselves about whether certain conditions should be classified and treated as autoimmune conditions. The list of "accepted" autoimmune disorders numbers several dozen, some of which we mention in other chapters of this book.

## Allergy

An allergic reaction is acquired (brought about by exposure to a triggering substance called an *allergen*) and rapid. Mild allergies are common in all human populations. Severe allergic reactions can be life-threatening.

Exposure to the allergen causes IgE antibodies to set off an excessive activation of certain types of WBCs (mast cells and basophils), which release excessive histamine. Histamine causes swelling of mucous membranes, such as in the nose and throat. The swelling causes nasal congestion and that annoying itch in the throat. Congestion and swelling can trap bacteria in the nasal cavities and lead to sinus infections or ear infections. The inflammatory response can lead to such symptoms as eczema, hives, and hay fever. Allergies are a significant factor in asthma.

*Anaphylaxis* is a rapid, severe, whole-body allergic reaction. Anaphylaxis results from an acquired hypersensitivity to an allergen. An initial exposure, called the *sensitizing dose,* to a substance like bee sting toxin or a protein in a food, produces no symptoms but sensitizes the person's immune system to the allergen. A subsequent exposure, called a *shocking dose,* sets off anaphylaxis. *Anaphylactic shock* is anaphylaxis associated with vasodilation (dilation of the blood vessels) over the whole body, which results in low blood pressure and severe broncho-constriction to the point where the person has difficulty breathing. Respiratory failure, shock, and heart arrhythmias can lead rapidly to death. The treatment is immediate administration of epinephrine. (Turn to Chapter 8 for information about epinephrine.)

## Chronic inflammation

The inflammation response is an important mechanism in the body's natural defense system against infection and disease. *Chronic inflammation,* on the other hand, is a disease. In chronic inflammation, the mechanisms that destroy invaders are turned on the body's own tissues.

Even low-level inflammation, such as occurs in a moderate case of gingivitis, can cause problems, and not just at the inflammation site. The proteins of the inflammatory response and the complement system can travel in the blood and harm cells and tissues anywhere in the body. Chronic inflammation is now widely recognized as an underlying disorder that contributes to many and diverse conditions, including cardiovascular disease; neurological diseases such as clinical depression and Alzheimer's; diabetes; many kinds of cancer; and even premature labor and preterm birth. The list of diseases and disorders that are now recognized to have an inflammatory component is getting longer all the time.

# Infectious diseases

Some microbes not only commandeer your body for their own purposes; they use your body as a platform from which to launch invasions into the bodies of all your closest associates. This is a brief look at two different types of chronic infectious viral diseases.

## HIV and AIDS

*HIV* is a species of virus that attacks cells of the human immune system, specifically the helper T cells. The immune system mounts a response, as it does to any other infection, and it may fight the virus to a standstill for years. But, as far as is known, the immune system is unable to eliminate the virus entirely. Like herpes viruses (see the next section), HIV hides out within cells. But as reactivated herpes viruses damage the nerve cells they hide out in, HIV damages the immune system cells, inhibiting the functioning of the immune system itself.

The diagnosis of *acquired immune deficiency syndrome* (AIDS) is made partly on the patient's status regarding certain infections that a healthy immune system has no difficulty beating back (opportunistic infections). Eventually, the response to pathogens and malignant cells is inadequate, and the patient succumbs to an opportunistic infection, cancer, or other disease.

## Herpes viruses

*Herpes* viruses are a leading cause of human viral disease, second only to influenza and cold viruses. People living to middle age usually have antibodies to most of the eight known human herpes viruses, whether or not they're aware of having been infected.

The immune system is capable of suppressing the herpes viruses but not of eliminating them. After a patient becomes infected by a herpes virus, the infection remains in the body for life. Following the primary infection, the virus may migrate to the *ganglia* (nerve bundles) and establish a latent infection, which may reactivate at any stage. Reactivation is frequently, but not always, associated with further disease. Immunocompromised patients are at risk for serious disease and death from reactivated herpes viruses, which are prominent among the opportunistic infections that have been the actual cause of death in many AIDS patients.

The *Varicella-Zoster virus*, the herpes virus that causes chickenpox, is usually acquired in childhood (unless the child has been vaccinated), and more than 90 percent of the population of the United States carries antibodies. The virus is spread by respiratory aerosols (inhaled particles) or by direct contact with the skin lesions of an actively infected patient.

For several days following initial infection, the virus sits in the respiratory mucosa, where it infects macrophages and lung cells. At this stage, no symptoms appear. The virus spreads to lymphocytes and monocytes, and then to epithelial sites throughout the body. The virus reaches the surface of the skin, and lesions, typically hundreds, form on the skin and mucosa, usually most pronounced on the face, scalp, and trunk. The disease is more severe in older children and adults and can be very severe in immune-compromised patients.

Reactivation of the virus may occur in late life. The recurrence of viral replication is accompanied by severe pain in areas innervated by the latently infected ganglia. Symptoms include chronic burning, itching pain and increased sensitivity to touch *(hyperesthesia)* called *post-herpetic neuralgia* or *shingles.* The pain may last months or years. Reactivation can affect the eye and the brain via certain cranial nerves.

# 5

# Life's Rich Pageant: Reproduction and Development

Chapter **14**

# The Reproductive System

This chapter is where you find out where babies come from and what happens when they're born. Like all animals, humans have an instinctive knowledge of mating. However, only we humans seem interested in understanding the processes of mating and reproduction. This chapter gives you information about the anatomy and physiology of reproduction. You'll have to look elsewhere for information on dating and mating rituals.

## Functions of the Reproductive System

The reproductive system is different from all the other body systems discussed so far. The other systems focus entirely on their own survival, but the reproductive system "risks it all" in an effort to contribute genes to future generations. It does nothing to enhance physiological well-being — in fact, it can pose severe threats to the organism's survival.

The following list gives you an overview of what the reproductive system is responsible for:

>> **Making gametes:** The *gametes,* also called *sex cells,* are made within the organs of the female and male reproductive systems. There are two kinds of gametes: The *ova* (singular, *ovum*) are the female gametes and the *sperm* (singular, *sperm*) are the male gametes. Specialized cells, called *germ cells,* generate the gametes in a cell-division process called *meiosis* (see the "Meiosis" section later in the chapter). At the cell level, the processes are essentially identical in female and male bodies. At the tissue, organ, system, and organism levels, the processes are very different. (We discuss the various processes throughout this chapter.)

TECHNICAL
STUFF

The terms *spermatozoon* (singular; literally "seed of an animal") and *spermatozoa* (plural) are the correct technical terms, used now almost exclusively in formal scientific writing.

>> **Moving gametes into place:** If the reproductive system is to succeed, one ovum and one sperm must make their way to the same place at the same time under the right conditions for them to fuse. Many of the reproductive system's tissues and organs chaperone the gametes from the place and time of their production to another place, where they're most likely to encounter their destiny.

>> **Gestating and giving birth:** Only the female reproductive system has organs for gestating a fetus and giving birth. (See the "Pausing for Pregnancy" section later in the chapter for a detailed discussion of pregnancy.)

>> **Nurturing the newborn:** The female reproductive system has tissues and organs specialized for nourishing the newborn for the first few months of life, until the older baby is capable of digesting other food.

# Producing Gametes

The process of meiosis includes the sequence of cell-level events that result in the formation of sex cells (gametes) from somatic cells (germ cells). (Flip to Chapter 3 for the details of cell division and differentiation.) Meiosis is the only cellular process in the human life cycle that produces *haploid* cells.

REMEMBER

Somatic cells are *diploid,* meaning that each cell nucleus contains two complete copies of the DNA that came into being in the zygote. Sex cells (gametes) are *haploid,* meaning that each cell nucleus contains only one copy of the DNA of the mother (somatic) cell. When two gametes fuse to form the zygote, each contributes its DNA to the new zygote, which is, therefore, diploid.

# Meiosis

All of our cells divide by *mitosis* for replacement, growth, development, and repair, as we discuss in Chapter 2. (Chapter 3 also has details about the cycle of cell growth and division.) Only a single designated cell type divides by *meiosis,* in order to produce gametes — that is, for purposes of sexual reproduction. The process of meiosis is similar in its mechanics to the process of mitosis, but there are several key distinctions between the two.

The most obvious difference is that meiosis has two parts, called *meiosis I* and *meiosis II.* Each part proceeds in a sequence of events similar to that of mitosis (prophase, metaphase, anaphase, and telophase). In mitosis, the mother cell is diploid, and both daughter cells are also diploid, each having one complete and identical copy of the mother cell's genome. In contrast, meiosis results in four haploid daughter cells. What's more, the four haploid genomes are all different.

The early stages of meiosis (prophase I in Figure 14-1) include a mechanism called *crossing-over* or *recombination* for exchanging genes between chromosomes. The result is that the cell that becomes the gamete (one of the four haploid products of meiosis) is carrying chromosomes that are completely unique and not identical to the mother cell's chromosomes. Like a lot of topics in cell biology, the complexity of these processes is beyond the scope of this book.

Note that replication of DNA does occur in meiosis, during the interphase that precedes the onset of meiosis I. After two sequential, dichotomous (in two) divisions, the two complete copies are distributed among the four daughter cells, each having received a single copy of each chromosome.

TECHNICAL
STUFF

Meiosis includes a number of mechanisms intended to ensure that each gamete has exactly one complete and correct copy of each gene. Any omission, duplication, or error is very likely to be fatal to the gamete or later to the embryo.

## Female gametes: Ova

A mature ovum (see Figure 14-2) is one of the largest cells in the human body, about 120 micrometers in diameter (about 25 times larger than sperm) and visible without magnification. The ovum contains a haploid nucleus, ample cytoplasm, and all the types of organelles usually found in the somatic cell, all within a plasma membrane. The plasma membrane is enclosed within a glycoprotein membrane called the *zona pellucida,* which protects the zygote and pre-embryo until implantation.

**1st Cell Division of Meiosis**

Parent cell

Prophase I

Metaphase I

Anaphase I

**2nd Cell Division of Meiosis**

Telophase I Prophase II

Metaphase II

Anaphase II

Telophase II

4 daughter cells

**FIGURE 14-1:** The process of meiosis.

Illustration by Kathryn Born, MA

Zona pellucida (glycoprotein membrane)

Sperm (25x smaller than secondary oocyte)

Female pronucleus (haploid)

Cell membrane (phospholipid bilayer)

Mitochondria (abundant)

Cytoplasm

Cumulus layer

Polar body developed during meiosis

Corona radiata cell

K·BORN

*Drawing not to scale

**FIGURE 14-2:** The human ovum.

Illustration by Kathryn Born, MA

*Oogenesis* (the development of ova) in humans begins in embryonic and fetal development with specialized somatic cells called *oogonia.* Millions of these cells head down the path of meiosis, producing cells called *primary oocytes.* However, they are suspended in meiosis at the prophase I point until the female reaches puberty. By birth, the human female has only about 700,000 primary oocytes left.

After the onset of puberty, the primary oocyte resumes meiosis I, producing two cells, called a *secondary oocyte* and the *first polar body.* However, cytokinesis happens unevenly so most of the primary oocyte's cytoplasm moves to the secondary oocyte. The first polar body completes meiosis II, and its daughter cells degenerate. The secondary oocyte continues with meiosis II but then stops again, this time during metaphase II.

The cells released from the ovary at ovulation are secondary oocytes. If the secondary oocyte is not fertilized, it degenerates without completing meiosis II.

When (or if) a sperm initiates fertilization, the secondary oocyte immediately resumes meiosis II, producing the ovum (plus a second polar body, which degenerates). Following fertilization, the ovum contains the sperm nucleus, and after approximately 12 hours, the two haploid nuclei fuse, producing the zygote.

# Male gametes: Sperm

A mature sperm has three parts: a head that measures about 5 x 3 micrometers, containing a haploid nucleus; a short middle section; and a long flagellum. The sperm is adapted for traveling light — it has very little cytoplasm (see Figure 14-3). The head is covered by a structure called the *acrosome* that contains enzymes that break down the ovum's membrane to allow entry. The middle contains mitochondria and little else. Mitochondria produce the energy that fuels the sperm's highly active flagellum, which propels the sperm through the female reproductive tract.

The process of sperm development *(spermatogenesis)* from meiosis to maturation takes place inside the *testes.* Specialized cells called *spermatogonia* divide by mitosis to produce another generation of spermatogonia. Mature spermatogonia, called *primary spermatocytes,* divide by meiosis, producing four haploid gametes called *spermatids.*

Similar to the case with females, males are born with spermatogonia in their *seminiferous tubules,* which remain dormant until puberty. During puberty, hormonal mechanisms pull the spermatogonia out of dormancy.

acrosome ——

nucleus ——

head

midpiece ——

tail

**FIGURE 14-3:**
The human
sperm.

© *John Wiley & Sons, Inc.*

In contrast to oogenesis, which is cyclic, spermatogenesis is continuous beginning at puberty and continuing lifelong in most men. In contrast to the one-per-month gametogenesis in females, males produce astronomical numbers of sperm. Each ejaculation produces about 1 teaspoon of semen, which contains about 400 million sperm in a matrix of seminal fluid. Mature sperm can live in the epididymis and vas deferens for up to six weeks.

## Determining sex

An important difference between males and females is that in females, all chromosome pairs are made up of two identical-looking strands, whereby in males, the strands are different from each other. This difference is easily visible under a high-power microscope: One of the pair is "normal" length (about the same length as all the other chromosomes) and the other is markedly shorter than all other chromosomes. The first is called the X chromosome, and females have one set of them in all their somatic cells. The second is called the Y chromosome, and males have a mismatched pair (one X and one Y chromosome) in all their somatic cells. After meiosis in the female, all ova have one X chromosome. After meiosis in the male, each sperm has either an X or a Y. The fusion of an ovum with an X sperm produces a female (XX) zygote. The fusion of an ovum with a Y sperm produces a male (XY) zygote.

## ERRORS IN SEX CHROMOSOME DISTRIBUTION

Occasionally, cell division processes go awry. For example, two chromosomes get pulled to one side of the cell, leaving the other without a copy at all. Numerous genetic disorders are caused by just such an error, *Down syndrome* being the most notable (people affected by Down syndrome have three copies of chromosome 21). The sex chromosomes (X and Y) are not exempt from this potential error, leading to offspring with more or fewer than two copies.

In *Klinefelter syndrome,* men have an extra X chromosome in their cells (XXY). Affected boys develop normally until they reach puberty. Then, because the testes are underdeveloped, very little testosterone is produced. Males don't develop the secondary sex characteristics associated with maleness (increased muscle mass, growth of body hair), may develop breast tissue, and are usually infertile. The earlier the condition is diagnosed, and the earlier testosterone replacement can begin, the greater the chance that the boy will develop normally and may even be able to father children with assistive reproductive procedures.

It is also possible for an offspring to develop with only a single copy of the X chromosome. Because males only receive one in the first place, this seemingly wouldn't cause too much trouble. However, girls born with *Turner syndrome* need regular medical care throughout their entire lives. In addition to the lack of sexual development, there are numerous physical symptoms (short stature, broad chest, and so on), and learning disabilities are common. Hormone replacement therapy is a necessary part of treatment, but the effects are widespread across all organ systems.

# The Female Reproductive System

The female body is specialized for reproduction to a much greater extent than the male body. The "Reproductive System (Female and Male)" color plate in the center of the book shows the female reproductive system in detail. Following is a brief discussion of the organs of the female reproductive system.

## Organs of the female reproductive system

The organs of the female reproductive system are concentrated in the *pelvic cavity.* Many of the female reproductive organs are attached to the *broad ligament*, a sheet of tissue that supports the organs and connects the sides of the uterus to the walls and floor of the pelvis.

## Ovaries

The *ovaries* are two almond-shaped structures approximately 2 inches (5 centimeters) wide, one on each side of the pelvic cavity. They house groups of cells called *follicles.*

The ovaries are the primary sex organs because they're the site of *oogenesis,* the process of oocyte maturation. The ovaries also have a major role in endocrine signaling, especially the production and control of hormones related to sex and reproduction, namely estrogen and progesterone.

Beginning at the female's puberty, the process of ovulation begins. The primary oocytes that have been dormant in her ovaries since early in her fetal development are hormonally activated, and secondary oocytes are released at a rate of approximately one per month from *menarche* (the first menstrual period of her life) to *menopause* (the last) — that is, from her early teen years to her late 40s or early 50s. The human female ovulates about 400 times during her lifetime.

## Uterus

The *uterus* or *womb* nourishes and shelters the developing fetus during gestation. It's a muscular organ about the size and shape of an upside-down pear. The walls of the uterus are thick and capable of stretching as a fetus grows.

The lining of the uterus, called the *endometrium,* is built and broken down in the *menstrual cycle,* which we discuss in the section "Cycling approximately monthly" later in the chapter. A portion of the endometrium *(deciduas basalis)* becomes part of the placenta during pregnancy.

The *cervix* is a cylindrical muscular structure about 1 inch (2.5 centimeters) long that rests at the bottom of the uterus like a thimble. It controls the movement of biological fluids and other material (not to mention, occasionally, a baby) into and out of the uterus. Normally, the cervix is open ever so slightly to allow sperm to pass into the uterus. During childbirth, the cervix opens wide to allow the fetus to move out of the uterus.

## Uterine tubes

The *uterine tubes* run from the ovary to the uterus. They are not literally connected to the ovaries; they just kind of hang over them. At the ovary end the uterine tube expands into a funnel shape called the *infundibulum.* This branches into fingerlike structures called *fimbriae,* which guide the egg into the uterine tube, which transports it to the uterus. The process of fertilization usually occurs in the uterine tube.

## Vagina

The *vagina* is the part of the female body that receives the male penis during sexual intercourse and serves as a passageway for sperm to enter into the uterus and uterine tubes. The vagina is about 3 to 4 inches (8 to 10 centimeters) in length. The cervix marks the top of the vagina.

During childbirth, the vagina must accommodate the passage of a fetus weighing on average about 7 pounds (3 kilograms), so the vagina's walls are made of stretchy tissues — some fibrous, some muscular, and some erectile. In their normal state, the vagina's walls have many folds, much like the stomach's lining. When the vagina needs to stretch, the folds flatten out, providing more volume.

## Vulva

In females, the external genitalia comprise the *labia majora, labia minora,* and the *clitoris.* Together, these organs are called the *vulva.* The term *labia* (singular, *labium*) means "lips." The labia of the vulva are loose flaps of flesh, just like the lips of the mouth (called *labia mandibulare* and *labia maxillare,* by the way). The labia protect the vagina's opening and cover the pelvis's bony structures.

Here are some details about the three parts of the vulva:

>> **Labia majora:** These large folds of skin — one fold on each side — cover the smaller labia minora. The labia majora extend from the *mons pubis* (pubic mound) back toward the anus. The mons pubis contains fat deposits that cover the pubic bone. Following puberty, pubic hair covers the mons pubis and the labia majora.

>> **Labia minora:** These hairless folds of skin lie underneath the labia majora and cover the opening of the vagina. The labia minora are attached near the vaginal orifice (opening) and extend upward, forming the foreskin that covers the clitoris.

>> **Clitoris:** This part of the vulva located above the vagina's opening and above the urethra has a shaft and glans tip, just as a penis does, and it's extremely sensitive to sexual stimulation. The clitoris contains erectile tissues that fill with blood during sexual stimulation. Because the tissue of the labia minora cap the clitoris, the swelling and reddening is also obvious in the labia minora. Stimulation of the clitoris can lead to orgasm in the female. Although females don't ejaculate, females do experience a building and release of muscular tension. Female orgasm causes the muscle tissue that lines the vagina and uterus to contract, which helps to pull the sperm up through the reproductive tract.

## Breasts

All humans have *mammary glands,* but only females produce a substance we call *milk* for the nutrition of relatively helpless infants with high calorie requirements. Besides nutrition, breast milk boosts the infant's immune system.

The breast contains about two dozen lobules that are filled with *alveoli* that make and store milk. The milk is released into *lactiferous ducts,* which merge at the *nipple* (see Figure 14-4). During puberty, the lobules and ducts develop, and adipose tissue is deposited under the skin to protect the lobules and ducts and give shape to the breast. During pregnancy, hormones increase the number of milk-producing cells and increase the size of the lobules and ducts.

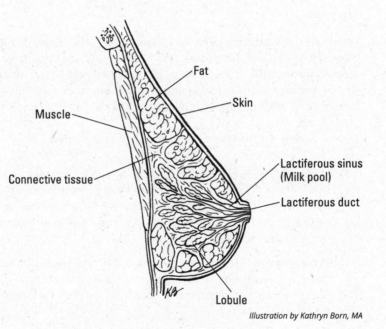

**FIGURE 14-4:** The human breast.

*Illustration by Kathryn Born, MA*

After the infant is born, the mother's pituitary gland secretes the hormone *prolactin,* which causes the milk-producing cells to create milk, and *lactation* begins. The infant suckles the milk out of the ducts through the nipple. Lactation continues as long as a child nurses regularly.

The hormone *oxytocin* is strongly involved in milk release (let-down reflex). Stimulation of the nipple prompts the secretion of oxytocin from the mother's pituitary gland. Oxytocin expels milk from the lobules by causing them to contract, just as it stimulates uterine contractions to expel the fetus. This hormone has also been strongly correlated with neuro-emotional phenomena, such as family bonding.

# Cycling approximately monthly

The *menstrual cycle (monthly cycle)* consists of both the *ovarian cycle* and the *uterine cycle*, both of which are approximately 28 days in duration (see Figure 14-5). These cycles run concurrently to prepare the ovum and the uterus, respectively, for pregnancy.

**FIGURE 14-5:**
The menstrual cycle.

By convention, the first day of menstrual bleeding is counted as Day 1 of the menstrual cycle. Menstrual bleeding begins at a point in the cycle when the levels of estrogen and progesterone are at their lowest. However, the entire menstrual cycle is directed by several hormones, not just estrogen and progesterone.

## Looking at the ovarian cycle

The 28-day ovarian cycle is the most important part of the menstrual cycle because it's responsible for producing the hormones that then control the uterine cycle. (See the next section, "Clocking the uterine cycle.") From Day 1 to Day 13, triggered by low estrogen level, *follicle-stimulating hormone* (FSH) stimulates the development of a follicle and *luteinizing hormone* (LH) stimulates the maturation of an oocyte in one of the ovaries. When the follicle is developed enough, it begins to secrete estrogen. When the level of estrogen reaches the appropriate level, a negative feedback mechanism involving the *hypothalamus* briefly slows the secretion of FSH and LH. When the follicle is fully mature and the oocyte is ready to be released, FSH and LH secretion spikes. This occurs on Day 14 and triggers *ovulation* (release of the oocyte). An oocyte lives for only 12 to 24 hours after ovulation (if unfertilized).

At the time of ovulation, the anterior pituitary gland, which has been secreting FSH and LH simultaneously, secretes a surge of LH that causes the follicle from which the oocyte was released to become a *corpus luteum* (yellow body). The corpus luteum secretes the hormone *progesterone*, which triggers the hypothalamus. When the corpus luteum has secreted a sufficient amount of progesterone, the hypothalamus stops the anterior pituitary gland from secreting any more LH. At that point, the corpus luteum begins to shrink (about Day 17). When the corpus luteum is gone (about Day 26), the levels of estrogen and progesterone are at the lowest levels of the cycle (sometimes causing symptoms of premenstrual syndrome), and menstruation starts (about Day 29, or Day 1 of the new cycle).

Like any cycle, the whole process starts over. When the level of estrogen is low during menstruation, the hypothalamus detects the low level and secretes *gonadotropin-releasing hormone* (GnRH), which prompts the anterior pituitary gland to release its gonadotropic hormone — FSH — so that another follicle is stimulated to develop a new oocyte that secretes estrogen. Now, you're back to the first paragraph of this section.

## Clocking the uterine cycle

The 28-day uterine cycle, which aims to prepare the uterus for a possible pregnancy, overlaps with the ovarian cycle.

>> **Days 1 to 5:** The first 5 days of the uterine cycle is when the level of estrogen and progesterone are lowest — the period of menstruation. The low level of sex hormones fails to prevent the tissues lining the uterus (the *endometrium*) from disintegrating and shedding. As the hormone levels drop, blood vessels spasm, cells undergo *autolysis* (self-destruction), tissues tear apart from the uterus wall, and blood vessels rupture, causing the bleeding that occurs during a period. The blood and tissue (menstrual flow) passes out of the uterus through the cervix and then out of the body through the vagina.

>> **Days 6 to 14:** During this *proliferative phase,* the developed follicle secretes high levels of estrogen, which makes the endometrium regenerate fresh tissue. The tissues lining the uterus and the glands in the uterine wall grow and develop an increased supply of blood. All these changes are preparation for nourishing an embryo and supporting a pregnancy, should the oocyte,

which is released on day 14, become fertilized and implant in the wall of the uterus. (See the section "Pausing for Pregnancy" later in this chapter.)

>> **Days 15 to 28:** During this *secretory phase,* the corpus luteum secretes an increasing level of progesterone, which further thickens the endometrium, and the glands of the uterus secrete a thick mucus. If the egg becomes fertilized, the thickened endometrium and mucus help to "trap" the fertilized egg so it implants properly in the uterus. If the egg doesn't become fertilized within a day or two, the corpus luteum begins to shrink because it won't be needed for a pregnancy. As the corpus luteum shrinks, the progesterone and estrogen levels decline, which causes the endometrium to "shred and shed" just before menstruation.

TECHNICAL
STUFF

Early on in a pregnancy, the corpus luteum serves as a source of estrogen and progesterone until the placenta develops and can secrete estrogen and progesterone on its own.

## Winding down the cycle

Physiologically, menopause essentially reverses the hormonal pathway of adolescence. When a woman enters menopause, her ability to reproduce ends — ovulation stops, and she no longer can become pregnant. She may also experience hot flashes and sweat baths if faulty signals from the parasympathetic nervous system disrupt the body's ability to accurately monitor its temperature. Other body processes slow down, including cellular metabolism and the replacement of the structural proteins in the skin, which leads to wrinkles. A woman's bones may also weaken when the breakdown of bone tissue occurs faster than the buildup of bone tissue during bone remodeling.

Menopause is one of the unique aspects of human physiology. Not that reproductive cycling declines and stops as the female ages. That happens in many mammal, bird, and reptile species (although relatively few individuals in any species live long enough to experience the decline of their reproductive capabilities).

The unique part is that human females often live a substantial proportion of their life span beyond their reproductive capacity. (A woman in her 80s has lived around 40 percent of her life after menopause.) Much research into this phenomenon is concentrated at the interface of biology and culture. One theory holds that an adult female with no offspring of her own to feed tends instead to feed her grandchildren or other children in her community. Children with grannies eat better, the theory goes, improving their chances of surviving to reproductive age and pushing her genes into one more generation.

# The Male Reproductive System

The male reproductive system produces sperm and moves them into the female reproductive system. On a rare occasion (relative to the astronomical number of sperm that the average human male produces), a sperm fertilizes an egg. All the billions and billions of other sperm a man produces in his lifetime have a limited life span — about six weeks if they stay in the male's body, or up to five days if in the female's body.

**REMEMBER**

The female gamete is released from the ovary as a secondary oocyte. Only when fertilization is initiated does the ovum (egg) come into being.

If *reproduction* is defined as ending with the creation of a new organism, that's the end of the male's reproductive function. If *reproduction* is defined as including the nurturance of the new organism until it is itself ready for reproduction, the anatomy and physiology of the male may well be substantially devoted to the task for decades, along with those of the female. For the purposes of this chapter, we use the more limited definition.

**TECHNICAL STUFF**

Some people are confused about the meaning of "evolutionary success" for the individual (you, for example). Truly, it's not how many zygotes arise from your gametes, or even how many offspring are born from these zygotes, but how many of these offspring survive to reproduce. Simply, evolutionary success is rated not by the number of your children (and certainly not by your level of sexual activity or number of opposite-sex partners) but by the number of your grandchildren.

## The organs of the male reproductive system

The organs of the male reproductive system produce gametes, called sperm, and transfer them to the female reproductive system. (Refer to the "Reproductive System (Female and Male)" color plate in the center of the book for a look at the male reproductive system.) In contrast to some other organ systems, and especially to the reproductive system of females, the male reproductive organs are located in an exposed location on the periphery of his body.

### Testes and scrotum

The *testes* (singular, *testis*) are paired organs that produce sperm and hormones. Like the ovaries of the female reproductive system, the testes are the site of gamete production and therefore the *primary sex organs*. Refer to Figure 14-6 for an illustration of the testicular structures.

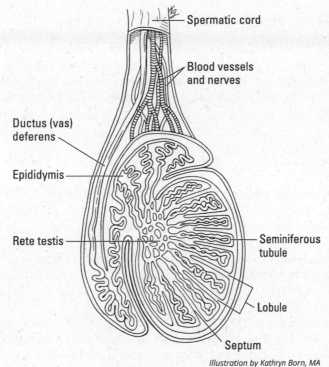

Spermatic cord

Blood vessels
and nerves

Ductus (vas)
deferens

Epididymis

Rete testis

Seminiferous
tubule

Lobule

Septum

**FIGURE 14-6:**
The testicular
structure.

*Illustration by Kathryn Born, MA*

The testes contain fibrous tissue that forms long, coiled compartments called *seminiferous tubules*. These are the site of *spermatogenesis*, the process of sperm development from meiosis. The walls of the seminiferous tubules are lined with thousands of *spermatogonia* (immature sperm). The seminiferous tubules also contain *Sertoli cells* that nourish the developing sperm and regulate how many of the spermatogonia are developing at any one time.

The seminiferous tubules transport the new sperm (spermatids) to another long, cordlike structure that lies atop each testis. This is the epididymis and maturation of the sperm occurs here. The *epididymis* is continuous with (that is, "becomes") the *vas deferens*, a tube that connects the epididymis of each testis to the penis. Sperm bide their time here until they're released via ejaculation.

The testes are held within the *scrotum* beneath and outside of the abdomen. The scrotum contains smooth muscle that contracts when the scrotal skin senses cold temperatures and pulls the scrotum (and thereby the testicles) closer to the body to keep the sperm at the right temperature. The scrotum's inside muscle layers are an outpouching of the pelvic cavity. The scrotum's outside skin is continuous with the skin of the perineum and groin.

### Prostate gland

Several other structures secrete the substances that make up the ejaculatory fluid that provides a matrix for the propulsion of sperm into the female reproductive tract. Among these are the *prostate* and the *seminal vesicles*. The prostate also contains some smooth muscles that help expel semen during ejaculation.

### Penis

The penis consists of a shaft and the *glans penis* (tip). The tube-shaped *urethra* runs through the shaft of the penis, and the *glans penis* contains the *urethral orifice*. The semen is ejaculated through the urethra and urethral orifice. *Foreskin* (also called *prepuce*) covers the glans penis. The foreskin of newborn males is often removed in a surgical procedure called *circumcision*.

During sexual arousal, erectile tissue in the shaft became engorged with blood, enabling the penis to be inserted into the female vagina, delivering the sperm to the vicinity of the secondary oocyte (if one is available).

**TECHNICAL STUFF**

The urethra and urethral orifice also function in the urinary system as the tube through which urine leaves the body. However, semen doesn't contain urine. At the point of ejaculation, a sphincter closes off the bladder to keep urine, which is acidic, from mixing with the sperm, which live in a basic environment.

## Seminal fluid and ejaculation

The *seminal vesicles*, glands located at the juncture of the bladder and vas deferens, have ducts that allow the fluid they produce to sweep the sperm from the vas deferens into the urethra.

Next, the *prostate gland* adds its fluid, which contains mainly citric acid and a variety of enzymes that keep semen liquefied. The prostate gland surrounds the urethra just below where it exits from the urinary bladder.

These two glands — the seminal vesicles and the prostate gland — secrete fluids that have several functions:

>> They're slightly basic with a pH of 7.5, just the way sperm like their environment to be.

>> They nourish the sperm by providing the sugar fructose so that the sperm's mitochondria can make enough energy to move its tail and travel all the way to the egg.

>> They contain *prostaglandins,* which are chemicals that make the uterus reverse its downward contractions. When the uterus contracts, the sperm are pulled farther upward into the female's reproductive tract.

As the glands add the secretions, forming the semen, pressure builds up on the structures of the male reproductive tract. When the pressure has reached its peak, the semen is expelled out of the urethra through the penis. Peristaltic waves (like those that occur in the digestive tract; see Chapter 11) and rhythmic contractions move the sperm through the vas deferens and urethra. The term for this discharge is *ejaculation* — part of orgasm in males, as is the contraction and relaxation of skeletal muscles at the base of the penis. As the muscles contract rhythmically, the semen comes out in spurts.

The *bulbourethral glands,* also called *Cowper's glands,* sit within the floor of the pelvis near the base of the penis on either side of the urethra. These two small glands have ducts leading directly to the urethra and secrete a mucuslike fluid in response to sexual stimulation. This clears the urethra of any acidity as well as provides lubrication for intercourse. Most of the lubrication, however, comes from the *vestibular glands* of the female, located near the vaginal opening.

REMEMBER

Sperm develop *motility* (the ability to move) in the epididymis, but they don't have the capacity to move until after ejaculation. That is, they don't actually "swim" until they have been released into the female reproductive tract. Prior to that, they're moved along by the ciliated cells that line the tubes.

# Pausing for Pregnancy

Pregnancy is established in two stages: fertilization of the secondary oocyte and implantation of the blastocyst in the uterus. Development of the embryo after implantation is the subject of Chapter 15. The female body makes many and various adaptations to pregnancy and delivery, which we examine in the following sections.

## Steps to fertilization

Ovulation sends a secondary oocyte from the follicle of the ovary into the uterine tube. Then, within an appropriate time, heterosexual intercourse results in the ejaculation of semen into the vagina. Some few million sperm make their way through the cervix, up through the uterus, and into the uterine tube to the vicinity of the waiting secondary oocyte.

To achieve fertilization, one sperm must penetrate the secondary oocyte's membrane, and its nucleus must fuse with that of the ovum. At this point, the secondary oocyte is fertilized, the ovum is developed, and the zygote comes into being (see Chapter 15).

The probability of any act of intercourse resulting in fertilization is actually quite low because many complicating factors exist. The timing of intercourse relative to ovulation is crucial. The released secondary oocyte is viable for only a matter of hours; sperm live a little longer in the female reproductive tract (one to two days on average). The environment within the female reproductive tract may be more or less hospitable to the sperm, depending on the female's hormone levels and other physiological processes. Even when a single sperm has made contact with the secondary oocyte, fertilization is not assured.

## Implantation

Following fertilization, the zygote divides immediately. Several more cell division cycles take place as the pre-embryo moves down the uterine tube. Experts believe that many pre-embryos die at this stage, sometimes because of genetic or developmental abnormalities. Only if the pre-embryo arrives at the uterus and properly embeds itself into the endometrium is pregnancy established.

A successfully implanted pre-embryo, now called a *blastocyst*, begins immediately to take over its mother's body. As the outer layer of the blastocyst begins to form the placenta, a hormone called *human chorionic gonadotropin* (hCG) is released, which maintains the *corpus luteum*, elevating levels of progesterone and estrogen and inhibiting menstruation.

**TECHNICAL STUFF**

The presence of hCG can be detected chemically in the urine of a pregnant woman within 10 to 14 days after fertilization. It may be detected before that symptomatically by the mother as a sensation of being nauseated in every cell of her body.

## Adapting to pregnancy

The maternal body responds to pregnancy with many anatomical and physiological changes to accommodate the growth and development of the fetus. Most structures and processes revert to the nonpregnant form (more or less) after the end of the pregnancy. See Chapter 15 for details about how the fetus grows in the uterus.

### Uterus

During pregnancy, the uterus grows to about five times its nonpregnant size and weight to accommodate not only the fetus but also the placenta, the umbilical cord, about a quart of amniotic fluid, and the fetal membranes. The size of the uterus usually reaches its peak at about 38 weeks gestation. During the last few weeks of pregnancy, the uterus has expanded to fill the abdominal cavity all the way up to the ribs. The size of the expanded uterus and the pressure of the full-grown fetus may make things difficult for the mother.

The placenta acts as a temporary endocrine gland during pregnancy, producing large amounts of estrogen and progesterone by 10 to 12 weeks. It serves to maintain the growth of the uterus, helps to control uterine activity, and is responsible for many of the changes in the maternal body.

Near the end of pregnancy, the cervix softens. Enlarged and active mucus glands in the cervix produce the *operculum*, a mucus "plug" that protects the fetus and fetal membranes from infection. The mucus plug is expelled at the end of the pregnancy. Additional changes and softening of the cervix occur at the onset of labor.

## Ovaries

Hormonal mechanisms prevent follicle development and ovulation in the ovaries.

## Breasts

The breasts usually increase in size as pregnancy progresses and may feel inflamed or tender. The areolas of the nipples enlarge and darken. The areola's *sebaceous glands* enlarge and tend to protrude. By the 16th week (second trimester), the breasts begin to produce *colostrum*, the precursor of breast milk.

## Other organ systems

Pregnancy affects all organ systems as they support the growth and development of the fetus and maintain homeostasis in the female. Here are a few important physiological consequences of pregnancy:

>> The other abdominal organs are displaced to the sides as the uterus grows.

>> Decreased tone and mobility of smooth muscles slows peristalsis and enhances the absorption of nutrients. An increase in water uptake from the large intestines increases the risk for constipation. Relaxation of the cardiac sphincter may increase regurgitation and heartburn. Nausea and other gastric discomforts are common.

>> Increases occur in blood volume, cardiac output, body core temperature, respiration rate, urine volume, and output from sweat glands.

>> Immunity is partially suppressed.

>> Spinal curvature is realigned to counterbalance the growing uterus. Slight relaxation and increased mobility of the pelvic joints prepares the pelvis for the passage of the infant. This can compromise the woman's lower-body strength starting in the second trimester.

# Labor and delivery

Labor is initiated by complex hormonal signaling between the maternal and fetal bodies. In the ideal labor and delivery process, called *parturition*, powerful contractions of the uterus push the fully mature fetus (infant) past the cervix and down the birth canal without undue trauma to either the mother or the infant. In this positive feedback mechanism (see Chapter 2), more stretch in the uterus and cervix triggers the release of hormones (particularly oxytocin), which cause stronger contraction and, thus, more stretch. Parturition occurs in three stages.

## Stage 1

Stage 1 labor progresses in three phases: early, active, and transition. Contractions intensify and the amniotic membrane ruptures (water breaks) at any point in this stage.

*Early labor* beings with irregular, mild contractions lasting around 30 seconds with 5 to 30 minutes between them. During the 8 to 12 hours of early labor, the cervix will become thinner, or *efface* (see Figure 14-7). It will also begin to dilate.

**FIGURE 14-7:** Effacement of the cervix in early labor.

*Illustration by Kathryn Born, MA*

After the cervix dilates to about 3 cm, *active labor* begins. Here, the contractions become more painful, squeezing inward from all directions. The contractions continually become stronger, last longer (45 to 60 seconds), and occur more frequently (every three to five minutes). Active labor generally lasts around three to five hours, until the cervix reaches about 8 cm.

During *transition,* the contractions shift from squeezing in to pushing down. This phase lasts only 30 minutes to two hours. The intense contractions last 60 to 90 seconds with only 30 seconds to 2 minutes between. Many women (who don't opt for an *epidural* to block all this pain) state that transition is the most painful part of the entire parturition process. On completion of this phase, the cervix is dilated to 10 cm and stage 2 begins.

## Stage 2

As the contractions continue at a now regular pace (60 to 90 seconds every three to five minutes), there is a natural urge to push. As the mother pushes with each contraction, the baby makes specific motions to ease its movement through the birth canal (see Figure 14-8). It flexes its head out, and then turns to aid passage of the shoulders.

a. Dilation of the cervix and breaking of amniotic sac

b. Delivery of the head

c. Delivery of the body

d. Delivery of the placenta

**FIGURE 14-8:** An overview of delivery.

*Illustration by Kathryn Born, MA*

This stage lasts about 30 minutes to two hours. Just after delivery, the infant's umbilical cord is cut and tied off. The infant is now totally separated from the mother and will soon have a stylish belly-button.

### Stage 3

After the baby is delivered, uterine contractions continue so that the placenta separates from the uterus wall. About 15 minutes after the baby is born, the placenta passes through the birth canal. Uterine contractions continue, during which the uterus contracts, returning eventually to near prepregnant size.

# Pathophysiology of the Reproductive System

Reproduction is a dangerous business for all animals. The investment of energy is huge, the risks are just as great, and the rewards are distant. It is an intricate interaction among many structures and processes besides those explicitly identified with the reproductive system. Structural defects, hormonal problems, genetic abnormalities, and cancers can all cause problems in reproduction. Here are some of the problems that can affect the reproductive system.

## Infertility

*Infertility* is the inability to fertilize or be fertilized. Infertility may be due to the failure to generate viable gametes, blockages in the "travel routes" of the gametes or the pre-embryo, or damage to or disease in the endocrine glands that control reproduction. Certain bacterial or viral infections, such as mumps, can result in *orchitis* (inflammation of the testes) that can affect fertility. Fertility decreases with age, ending abruptly at menopause in females and declining more gradually in males. Turn to Chapter 15 for more information on aging.

## Sexually transmitted infections

Some microbial diseases are transmitted via sexual contact. These *sexually transmitted infections* (STIs) are endemic in human populations (that is, they exist in all human populations to one extent or another and have since forever) because bacteria and viruses can easily be spread from one person to another via the organs and secretions of the reproductive system.

The terms STI and STD (sexually transmitted disease) are often used interchangeably, but STI is becoming more prevalent. The reason for the shift is that STD implies that the presence of the pathogen led to characteristic symptoms. This, however, is not always the case. The microbe can be passed on without any indication of disease (hence, STI).

Sexually transmitted infections are similar in most ways to other microbial infections. All the major groups of microbes (bacteria, fungi, protists, and viruses) have evolved some ability to propagate themselves in the very hospitable environment of the human reproductive tract. They cause problems in all the same ways: by inducing inflammation, over-activating the immune response, and destroying cells.

In the very special case of *HIV*, the infectious organism (an exotic creature called a *retrovirus*) destroys structures of the immune system, which leaves the body vulnerable to attack by other microbes.

The infectious bacterium *Chlamydia trachomatis* is transmitted by other means as well as sexually; it infects the human eye, joints, and lymph nodes, and also takes up residence in arteries.

## Premenstrual syndromes

Up to 80 percent of menstruating women experience both physical and mental changes just before the onset of menstrual blood flow. The type and severity of symptoms varies greatly from one woman to another, but the symptoms tend to remain stable through her reproductive life. *Premenstrual syndrome* (PMS) is a term used to describe a medical syndrome of mood swings, mild *edema* (fluid retention in the tissues), irritability, fatigue, food cravings, and uterine spasms (cramps) that affect an estimated 20 percent to 40 percent of women. Another more severe form, called *premenstrual dysphoric disorder* (PMDD), affects 2 percent to 10 percent of menstruating women, has much in common with mood disorders, and can result in severe disruption of daily activities. Physiological, psychological, environmental, and social factors all seem to play a part in the development of these disorders.

## Endometriosis

The *endometrium* is the lining of the uterus, which is shed during menstruation. In *endometriosis*, endometrial tissue grows in or on organs of the body other than the uterus — usually organs in the pelvic cavity, such as the bladder, ovary, or large intestine. Because the ovaries are not directly attached to the uterine tube, the endometrial tissue migrates and "falls out" into the pelvic cavity. During the uterine cycle, the tissue, regardless of its location, responds by building up and then disintegrating, sometimes causing extreme pain.

## Cryptorchidism

During fetal development, the testes are located inside the pelvic cavity, but around the time of birth, the testes descend into the scrotum. Failure of testes to descend is called *cryptorchidism.* Unless corrected by surgery, cryptorchidism results in sterility.

## Hypogonadism

Problems with the pituitary gland, such as injury or tumors, can cause *hypogonadism,* a decline in the function of ovaries or testes. The pituitary gland secretes follicle-stimulating hormone (FSH), which normally spurs on maturation of the oocyte or spermatocyte and the subsequent release of estrogen or testosterone. Symptoms in women include *amenorrhea* (absence of menstruation) and infertility. In men, the symptoms of hypogonadism are impotence and infertility.

## Erectile dysfunction

*Erectile dysfunction* (ED), also called *impotence,* is a condition in which penile erection doesn't follow sexual stimulation. ED has a variety of possible causes, including damaged blood vessels, sometimes because of diseases such as diabetes; psychological factors such as stress and fear; and nerve damage. Some degree of ED is considered a normal part of aging.

## Pathophysiology of pregnancy

Even a healthy, young-adult woman carrying a pregnancy under ideal social and economic conditions is at risk for pathophysiology in many organ systems. Carrying a pregnancy and giving birth are major contributors to disability, disease, and death in females. And her existing dependent children experience consequences as well.

Childbirth has other risks — the wounds incurred expose a woman to some kinds of infections, and significant blood loss is quite common. Some other pregnancy-related disorders are

>> **Ectopic pregnancy:** An *ectopic pregnancy* is an abnormal pregnancy where the pre-embryo implants outside the uterus, most commonly in the uterine tubes. An ectopic pregnancy is often caused by a condition that blocks or slows the movement of a pre-embryo through the uterine tube to the uterus. This may be caused by a physical blockage, hormonal factors, and by other factors, such as smoking. The fetus can't survive and often stops developing altogether. Ectopic pregnancy is a life-threatening condition.

>> **Gestational diabetes:** Gestational diabetes is *hyperglycemia* (excess glucose in the blood) that develops during pregnancy and that affects both mother and fetus. Diabetes can complicate delivery and can increase risk for diabetes mellitus after the pregnancy.

>> **Incompetent cervix:** In this condition, the cervix is unable to support the pregnancy. The cervix may have been traumatized during an earlier birth. A woman carrying multiples is at increased risk. The cervix dilates prematurely (before the onset of labor), a serious risk for pregnancy loss. To alleviate the problem, in a procedure called *cerclage,* the cervix can be stitched to give it extra support.

>> **Pre-eclampsia and eclampsia:** Just as urine is checked for glucose to help stave off gestational diabetes, the urine is also checked for protein, diagnostic of risk for pre-eclampsia. *Pre-eclampsia* (high blood pressure during pregnancy) can easily escalate into *eclampsia,* characterized by seizures, and possibly coma or even death.

>> **Placenta previa:** In this condition, the placenta covers the cervix, partially or completely, blocking delivery of the fetus or causing heavy bleeding. Placenta previa in the second or third trimester may result in blood loss from the placenta as the growing fetus presses on it. The bleeding puts the mother at risk for preterm labor and the fetus at risk for premature birth. Pregnancies complicated by placenta previa are treated by cesarean delivery.

>> **Placental abruption:** In this condition, part of the placenta peels away from the uterine wall before the fetus is ready to be delivered. Such an event can deprive the fetus of oxygen and nutrients and, depending on the extent of the tear, can bring on preterm labor and cause life-threatening hemorrhaging in the mother. Maternal hypertension, physical trauma to the mother and fetus (such as a car accident), and a short umbilical cord are among the most common causes of placental abruption.

>> **Fetal distress:** Labor and delivery are stressful for the fetus. Fetal distress can sometimes complicate the delivery.

# Pregnancy loss

Pregnancy loss, also called *spontaneous abortion* and *miscarriage,* is the death of an embryo or fetus without apparent cause within the first 20 weeks of gestation. From 10 percent to 25 percent of all clinically recognized pregnancies end in miscarriage. Many more pregnancies end shortly after implantation, often without the woman realizing that she was pregnant. The bleeding that results may begin around the expected time of her menstrual period.

Miscarriage has many causes. Clinicians speculate that many may be caused by abnormalities in the embryo and not because of any disorder of the woman's reproductive system. Many women experience miscarriage and go on to have a normal pregnancy with a normal outcome (a cute baby). Repeated miscarriage, however, may indicate some form of disorder in either the male or female parent.

# Chapter **15**

# Change and Development over the Life Span

In the context of anatomy and physiology, *development* means the pattern of change through an organism's lifetime. Development is closely related to the specialized branch of biology called *ontogeny*, which studies an organism's history within its lifetime. Human development has been a subject of much study for thousands of years, especially for parents and grandparents of young human children. It's a good thing for babies that people find them so fascinating: The human baby takes a lot of effort to maintain and a long time to mature.

In this chapter, we take a look at human development, from the creation of the zygote through old age. Although you've already experienced some of this development, you get a glimpse of some of the changes your body will go through as you age.

# Programming Development

One way to think of development is as the unfolding in real time and space of a program for generating a unique biological organism. The program is launched when a new *zygote* comes into existence. All zygotes are created the same way and then proceed down the path of development encoded in their own species-specific and individual-specific DNA. (Flip back to Chapter 14 if you need a refresher on the zygote.)

The totality of the DNA of a zygote — that is, its *genome* — comes into existence at the time of fertilization. The DNA in the zygote's nucleus comprises genes (specific DNA sequences) from both its parents, 50-50, but this particular combination of genes has never been seen before and will never be again.

Most genomes, including all human genomes, have aging and death built into the program. All die sooner or later. A few survive until their program has fully unfolded and reached its end.

## Stages of development

Development begins in the zygote and continues until death. There's no universally agreed-upon definition of the development stages (although two milestone events — birth and, for females, the onset of menstruation — are universally acknowledged), and the age range at which a person passes from one stage to the next is wide. Change is more or less continuous through life, and different organ systems undergo significant changes on their own development timetable. However, conventionally in human biology, the development milestones that mark the stages are based on developments in the nervous and reproductive systems.

## Dimensions of development

The structural and physiological changes that happen during human development include an increase in size, the acquisition of some specialized abilities, and the loss of some other specialized abilities continuously throughout life. When all goes well, *senescence* (aging) is the final stage of development.

The following sections assume an organism for whom all is going well, biologically speaking: no fatal errors in the genome itself and adequate resources to sustain nutrition, thermoregulation, and all the rest of the life-maintaining physiological reactions.

## Growth

Part of human development involves an increase in size. Increased size is primarily accomplished by the growth of organs that exist in some form in the embryo: The heart grows larger, the brain grows larger, and the bones get longer and heavier. The organs grow by building more of their own tissues, and tissues get bigger by adding cells or increasing cell size. Everything (well, almost everything — there are always exceptions in biology!) grows together, mostly by adding cells.

However, not everything grows equally. Different stages of development are characterized by different proportions of tissue types. For example, both the brain and the skeletal muscle increase in size from infancy to adulthood, but the proportion of muscle tissue to brain tissue is much higher in adulthood.

When a three-dimensional object such as a living body increases in size, the *surface-to-volume ratio* decreases. (Or, to put it another way, the *volume-to-surface ratio* increases — more of your inside parts are dependent on fewer of your outside parts to interact directly with your environment.) The size of a human body strongly influences thermoregulation, fluid balance, and other key aspects of homeostasis.

## Differentiation

For humans, the acquisition of new abilities or improvements in existing abilities is part of development. New physiological abilities come about usually because of cell and tissue *differentiation* (function specialization). Tissue specialization begins in the pre-embryonic stage, as we discuss in the "Development before Birth" section later in this chapter. A newborn has some version of more or less all the cell and tissue types, but many fully differentiated cells must be generated and integrated functionally into tissues at appropriate stages of development. See Chapter 5 for a brief description of the development of a *long bone* (an organ of the skeletal system) by cell growth and differentiation.

Lots of human body functions aren't "learned" but "developed." The ability to digest starch, for example, is acquired during the first year of life, when the body starts producing the necessary enzymes — not when someone teaches a baby how. Toilet training is more about the maturity of the nervous system than the diligence of the parents. The acquisition of a new skill, structure, or process is sometimes accompanied by the loss of existing abilities. A young adult is better at planning than a teenager but has most likely lost some stamina for all-nighters, parties, and road trips. The stages of human development can be characterized by these abilities gained and lost.

Among many aspects of development research, brain research has yielded very interesting data in recent years, aided by advanced imaging technology (see Chapter 1). In the late 1990s, the decades-old doctrine that humans don't generate any new brain cells after birth was definitively shown to be false. Data from many different kinds of studies since then have indicated that the human brain is *plastic* (capable of change and development) well into old age. Turn to Chapter 7 for more about brain development.

## Senescence

According to recent theories, age-related decline in specialized and even basic physiological functions is built into new genomes right at the start. Structures at the ends of the chromosomes called *telomeres*, which get shorter and shorter as a genome ages, control the number of times the genome can replicate. Gradually, cells lose the ability to divide. The number of aged and dying cells in a tissue eventually exceeds the number of new cells of their type being made to replace them. The tissue loses its ability to function, which impairs the organism's survival.

The aging processes are an active area of research in anatomy and physiology. In recent decades, therapies and devices to counter aging's effects have dominated the medical products marketplace worldwide.

# Development before Birth

Human birth is a commonplace miracle: from a single infinitesimal cell to a human baby in less than ten months. The following sections give a brief overview of how it happens.

If you want more detailed information on pregnancy, check out *Pregnancy For Dummies,* 3rd Edition, by Joanne Stone, Keith Eddleman, and Mary Duenwald (Wiley).

## Free-floating zygote to protected embryo

Chapter 14 covers the events leading to the fertilization of a secondary oocyte and implantation of the blastocyst in the uterus from the point of view of female reproductive anatomy and physiology. This section covers those same events from the zygote's point of view, from the fusion of the haploid genomes of the parent gametes to implantation in the uterus (see Figure 15-1).

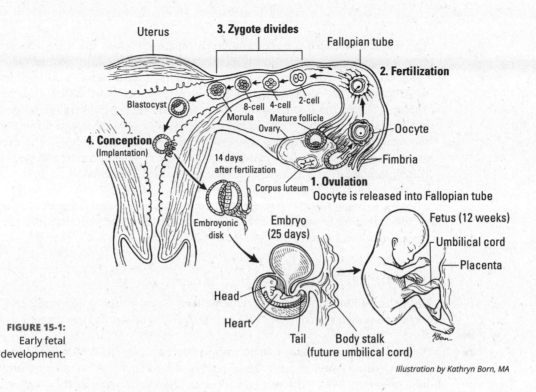

**FIGURE 15-1:**
Early fetal
development.

*Illustration by Kathryn Born, MA*

## Beginning it all

Fertilization, which takes about a day, begins when a sperm penetrates a *secondary oocyte* (an egg). After a sperm binds with the receptors in the *zona pellucida* (see Figure 14-2), it uses enzymes in its acrosome to digest the egg's protective layer. When the sperm has finally reached the cell membrane of the oocyte, it attaches to receptors there. This triggers two important events:

» **The zona pellucida will harden, preventing another sperm from fusing with the egg's actual cell membrane.**

» **The oocyte will restart meiosis II.** As this happens, the sperm's nucleus is allowed into the cell. This way, when the nucleus is reforming in telophase II, the DNA contribution of the sperm is incorporated. The cell, now officially considered a zygote, has completed fertilization and ready to begin its journey.

## A dangerous journey

The zygote undergoes *cleavage* (mitotic division) immediately. Over the next few days, the daughter cells (called *blastomeres*) divide twice more, to a total of 16 blastomeres, all within the rigid wall of the zona pellucida with no increase in overall size.

The mass, now called a *morula* (mulberry-shaped), leaves the uterine tube and enters the uterine cavity. Cell division continues, still confined within the zona pellucida, and a cavity known as a *blastocoel* forms in the morula's center. Around the sixth day after fertilization, the hollow structure, now called a *blastocyst*, "hatches" from the slowly eroded zona pellucida within the uterine cavity. The outer layer of blastocyst cells secretes an enzyme that facilitates implantation in the endometrium. *Angiogenesis* (building of blood vessels) begins in the uterus, and diffusion between mother and blastocyst begins. When this diffusion is established, implantation is complete and the pregnancy is established.

The new genome has survived a very dangerous stage of development. Biologists estimate that up to one-half of blastocysts fail to implant, and they die. But the new genome still has challenges ahead.

## The embryonic stage

Weeks three through eight after implantation are called the *embryonic stage*. During these weeks, the embryo's cells begin to differentiate and specialize.

The transition from blastocyst to *embryo* begins when the implanted blastocyst develops into a two-layer disc. The top layer of cells *(epiblast)* becomes the embryo and amniotic cavity; the lower layer of cells *(hypoblast)* becomes the yolk sac that nourishes the embryo. A narrow line of cells on the epiblast, called the *primitive streak,* signals *gastrulation* — cells migrate from the epiblast's outer edges into the primitive streak and downward, creating a new, middle layer. By 14 days or so after fertilization, the embryo, now called a *gastrula,* has *ectoderm, mesoderm,* and *endoderm* layers — the very beginning of tissue formation.

## The fetal stage

Following week eight and lasting until birth is the *fetal stage.* Growth and development occur rapidly during this time. The *primary germ layers* continue their development as the fetus becomes more recognizable as a baby. The ectoderm develops into the skin and nervous tissues, while the endoderm forms your inner tubes — the alimentary canal and the respiratory tract. The mesoderm develops into everything in between, including the bones and muscles.

The major milestones of fetal development are discussed later in the "Dividing development into trimesters" section.

## Forming the placenta

Immediately after implantation, the blastocyst initiates the formation of the *placenta,* a special organ that exists only during pregnancy that's made of the mother's cells in the outer layers and the fetus's cells in the inner layer.

The placenta serves to support the sharing of physiological functions between the mother and the fetus: nourishment (provision of energy and nutrients), gas exchange (a fetus must take in oxygen and eliminate carbon dioxide before birth), and the elimination of metabolic waste. The placenta allows some substances to enter the fetal body and blocks others. It does a good job of delivering nutrients and maintaining fluid balance, but it's permeable to alcohol, many drugs, and some toxic substances.

The placenta is a dark red disc of tissue about 9 inches (23 centimeters) in diameter and 1 inch (2.5 centimeters) thick in the center, and it weighs about a pound (roughly half a kilogram). It connects to the fetus by an *umbilical cord* of approximately 22 to 24 inches (56 to 61 centimeters) in length that contains two arteries and one vein. The placenta grows along with the fetus.

Nutrients and oxygen diffuse through the placenta, and the fetal blood picks them up and carries them through the umbilical cord. Then, the wastes that result from the fetus metabolizing the nutrients and oxygen are carried back out through the umbilical cord and diffused into the placenta. The mother's blood picks up the wastes from the placenta, and her body excretes them. Geez, moms start cleaning up after their kids before they're even born!

Both the fetus and the placenta are enclosed within the *amniotic sac,* a double-membrane structure filled with a fluid matrix called *amniotic fluid.* The fluid keeps the temperature constant for the developing fetus, allows for movement, and absorbs the shock from the mother's movements.

## Dividing development into trimesters

Officially, by convention, Day 1 of a pregnancy is the first day of the woman's previous menstrual period. Obviously, she wasn't pregnant on that day, nor for numerous days thereafter. But it's easier to be sure about the start date of a menstrual period than about the day of ovulation or fertilization or implantation, so that's the custom that doctors follow. Then, by convention, doctors count ahead 280 days to arrive at the *due date,* the date on which, if all pregnancies and all babies were alike, the birth would take place. The 280 days, usually expressed as 40 weeks, are the human *gestational period* (length of pregnancy). This period is divided, again by convention, into three trimesters, though nothing specific marks the transition from one to the next.

The following sections provide an overview of the development of a fetus's organs through the three stages of pregnancy. See the "Prenatal Development" color plate in the center of the book to get an idea of what a developing fetus looks like.

## The first trimester

All the body's organs begin development in the first trimester. The cardiovascular system forms from small vessels in the placenta three weeks after fertilization. The heart begins to beat at this time as well.

During the second month, the organ systems continue to develop, and the limbs, fingers, and toes begin to form. The embryo starts to move at the end of the second month, although it's still too small for the mother to feel its movements. Also during the second month, ears, eyes, and genitalia appear, and the embryo loses its tail and begins to look less like a sea horse and more like a human.

At the end of the first trimester, the fetus is about 4 inches long (10 centimeters) and weighs about an ounce (28 grams). The head is large, and hair has begun to grow. The intestines are inside the abdomen, and the urinary system (kidneys and bladder) starts to work.

TIP

If you're counting weeks and feel you're losing track, remember that the trimesters of pregnancy are measured from Day 1 of the mother's last menstrual period. This date is around two weeks earlier than the date of fertilization. The embryonic stage is the second and third month of pregnancy.

## The second trimester

The fetus, with all its systems in place, continues programmed development in the second trimester. *Ultrasound imaging* shows the skeleton, head details, and external genitalia. Bone begins to replace the cartilage that formed during the embryonic stage. At the end of the second trimester, the fetus is about 12 to 14 inches (30 to 36 centimeters) long and weighs about 3 pounds (1.4 kilograms).

## The third trimester

The fetal development program speeds up in the third trimester. The fetus, with its systems developed, continues to grow in size. Subcutaneous fat is deposited, which serves as a critical energy reserve for brain and nervous system development.

Near the end of the third trimester, the fetus positions itself for birth, turns its head down, and aims for the exit. When the fetus's head reaches the *ischial spines* of the pelvic bones (see Chapter 5), the fetus is said to be *engaged* for birth. (See Figure 15-2.)

**FIGURE 15-2:**
A fetus late in the
third trimester.

*Illustration by Kathryn Born, MA*

# The Human Life Span

Mammals have a general pattern of development from birth to senescence, and humans closely follow that pattern in most ways. In addition to following a typical sequence of events (live birth of dependent young, late development of the reproductive system, hairiness increasing with age, and so on), the mammalian pattern has some "rules" that tie the size of the animal with the pace of development. Generally, the larger the mammal (typical adult size range), the longer the development period. Humans are right where you would expect them to be on this curve.

Every species of mammal has a species-specific version of the pattern, of course, encoded in the species genome. The human species genome encodes a long infant dependency period and, for females, an extraordinary prolongation of life beyond her reproductive years.

## Changes at birth

The timing of birth is a compromise between the anatomical needs of a large brain and those of a bipedal gait. The fetus is born a little earlier than would probably be ideal from the point of view of its development, but the pelvis of the adult female has grown narrower and less flexible to support a redistribution of weight and bipedal mobility. Evolution has favored a compromise: a period of a few weeks when the baby, though fully separate from the mother, is still developing in ways that other mammals have completed before birth. Evolution continues to support this compromise relentlessly.

The newborn undergoes a number of changes at birth to allow it to survive outside the womb and adapt to life in a new cold, dry, and terribly immediate environment.

>> **The first breath:** The fetus exchanges oxygen and carbon dioxide through the placenta. At birth, the newborn's lungs are not inflated and contain some amniotic fluid. Within about ten seconds after delivery, the newborn's central nervous system reacts to the sudden change in temperature and environment by stimulating the first breath. The lungs inflate and begin working on their own as the fluid is absorbed by the lungs or coughed out.

>> **Thermoregulation:** Almost as quickly as the lungs start functioning, temperature receptors on the newborn's skin mediate the generation of metabolic heat by muscle action (shivering) and the burning of stores of *brown fat*. Very little brown fat is present in adults, but it's prevalent in infants, generating heat by "burning" the lipids stored in it.

>> **Digestive system:** The newborn's digestive system starts to work in a limited way immediately after birth. A newborn can digest colostrum and breast milk. Even so, the digestive system can take several weeks to settle down to efficient functioning.

In the fetus, the liver acts as a storage site for sugar (glycogen) and iron. After birth, the liver begins to takes on its other functions. It begins breaking down waste products such as excess red blood cells.

>> **Urinary system:** The fetus's kidneys begin producing urine by the end of the first trimester of pregnancy. The newborn usually urinates within the first 24 hours. The capabilities of the kidneys increase sharply through the first two weeks after birth. The kidneys gradually become able to maintain the body's fluid and electrolyte balance.

>> **Immunity:** The immune system begins to develop in the fetus and continues to mature through the child's first few years of life. The development of immunity is an active area of research; there is still much we don't understand. Mothers pass their antibodies on through fetal circulation and when nursing. Breast milk also contains components shown to promote the development of an infant's own immune system.

# Infancy and childhood

The long human infancy and childhood is one of the wonders of the biological world.

All organ systems grow and develop in infancy almost as rapidly as during fetal development. (All except the reproductive system. See the upcoming "Adolescence" section.) The physical development milestones in the first year alone would take a full book this size to describe. Just looking at a year-old infant and then at the infant's photos at birth shows you a lot.

Most infants double or triple their birth weight in the first year. Besides adding size to all tissues and organs, the new cells differentiate elaborately and add functionality, according to the individual's genomic scheme. The baby's skeleton changes size, proportion, and composition and takes on all the attributes of bipedalism. The mouth acquires the subtle muscle control to form words and kisses. The baby starts applying the characteristically human opposable thumb to everyday tasks (picking up toys and throwing them down again) in the second half of the first year. The brain grows in size and, just as important, the number and complexity of connections grow astronomically.

A toddler child (between ages 1 and 3) develops sphincter control. Social life begins. Vigorous play coordinates the development of the musculoskeletal and nervous systems. The mechanisms of homeostasis gradually strengthen.

Overall, from infancy to adolescence, the human child becomes bigger, stronger, and smarter every day. The child steadily gains control of the body on a conscious and physiological level. The child is typically fluent in at least one spoken language by age 6. A fully human degree of hypersociability is frequently evident in preadolescents (about age 10 to puberty).

And yet the juvenile's dependency goes on and on. Although most children are weaned before age 2 and can walk fairly long distances by age 10, they are, for the most part, hopeless at obtaining food and shelter. Their caregivers do that for them.

Their physical and mental development is devoted, instead, to mastering the unique aspect of human life called "culture." That takes these brilliant human children about 20 years of intense study. During the period of human evolution, the survival of any individual was dependent primarily on the survival of the individual's kin group. Participating effectively in the culture has always been the best way for humans to increase the likelihood of their own survival and the survival of those who carry their genes.

# Adolescence

One organ system doesn't undergo much development in infancy and childhood: the reproductive system. It remains in a state of suspended animation from early in fetal development until *puberty*, the first part of the development stage called *adolescence* (*adol-* means "adult").

During puberty, the reproductive systems of males and females emerge from the suspended state. This usually happens sometime between age 11 and 14 for girls; in boys, it begins a couple of years later. Puberty ends when the reproductive system is mature enough to produce viable gametes. (That is, reproduction becomes physically possible.) Hormones play a very large part in this complex process, and getting them all to function smoothly together typically takes a few years.

The hormones produced during adolescence cause physical and neurological changes in female and male bodies. The frequent hormone surges cause acne, among other miseries, and they often trigger emotionally uncomfortable mood swings. The brain's "executive" functions (judgment, impulse control, and risk-assessment) are frequently impaired.

Growth spurts are common during puberty, and growth continues through adolescence. Bones lengthen, muscle mass increases, and all organs reach near-adult size. The primary and secondary sex organs grow and mature. Fat and muscle are redistributed. Adolescents can maintain high levels of physical activity, fueled by the output of millions of new mitochondria. Adolescent sleep-wake cycles may be strikingly different from those of children and adults.

## Female puberty

In females, the ovarian and uterine cycles begin (see Chapter 14), which becomes evident when menstruation starts. After the female is ovulating, pregnancy is possible. The female breast develops during puberty.

Other changes that occur in females during puberty include growth of hair in the *axillary* (armpit) and *pubic regions,* and the development of the female fat distribution pattern: more on the hips, thighs, and breasts.

## Male puberty

Hormones from the *anterior pituitary* in boys allow them to produce testosterone, and as a result, to begin to develop sperm regularly. Testosterone has certain effects, such as initiating the growth of facial and chest hair, building lean muscle, causing hair to develop in the axillary and pubic regions, and making hair on arms and legs dark and coarse. The vocal cords thicken and lengthen, which causes deepening of the voice. The penis and testes enlarge. Males develop broader shoulders and narrower hips than females.

# ANATOMY AND CULTURE

Just about everything about human development is more or less "typical" for mammals: The young are born alive, ready to breathe air and suck milk from their mothers' mammary glands. As for size, human infants are within the very wide range of normal for a mammalian infant (between a shrew and a whale). Some mammalian infants are ready to run with their herd at one day of age, but lots of mammal babies are pretty helpless.

Only one thing is very odd about human development *(ontogeny),* and that's the exceptionally long duration of immaturity. Compared with any other known species, even humans' closest evolutionary relatives, human babies take a long time to grow up. The evolution of this remarkable trait has been the subject of much scientific research and speculation at least since Darwin in the fields of anthropology, evolutionary biology, psychology, and genetics.

The consensus is that the long human childhood extends the brain's development, giving it time to absorb the culture's complexities. Humans are a hypersocial species, and learning culture is an absolute requirement for survival. The period between weaning and adulthood is strongly devoted to learning spoken language, nonverbal communication, and other aspects of culture. The very food that humans eat must be prepared within the social group.

Anatomy and culture have been developing together for a few million years. All organ systems have adapted to culture, and the human animal is fully hypersocial. Humans have never been independent of the social group.

# Young adulthood

Typically, young adulthood is a time of good health and resilience. Many people complete their parents' bid for evolutionary success during this stage. (That is, they become parents themselves.) Energy levels can remain high all through the 30s, 40s, and 50s for some people. However, after growth is completed, energy requirements decrease, and adults must decrease their caloric intake to avoid accumulating body fat that can put long-term stress on several organ systems.

The gradual physical decline of senescence actually begins during these years. Influenced by genetic and environmental factors, arteries begin to accumulate damage, slightly more bone is lost than is made, and the same thing happens with the structural proteins of the skin. Muscle mass declines slowly but not imperceptibly. Damage accumulates from repetitive injuries, bad habits, and bad genes.

# Middle age

For most people, young adulthood ends sometime in the 40s or 50s. The cellular cycles that replace cells and repair tissues slow down. Loss of bone and muscle mass accelerates. However, most of the time, these losses aren't critical and can be mitigated by medical therapies or lifestyle adjustments (diet, exercise, sunscreen, and so on).

For men, reproductive capacity diminishes. For women, it disappears altogether at *menopause*, usually around age 50. Production of some hormones diminishes, triggering anatomical and physiological changes great and small.

The brain, however, continues to develop, cognitively and in many other ways. Some recent brain research shows that an older brain thinks better about some things than a younger brain does, including making financial decisions, exercising social judgment (intuitive judgments about whom to trust), and recognizing categories. The older brain is better at seeing the proverbial forest and following the gist of an argument. These are typically the peak years of professional or occupational achievement. In addition, across all occupations and ethnicities, a sense of well-being peaks as people reach middle age.

An extraordinary aspect of human development is the length of this period. Human females commonly outlive their fertility by three decades — a third of the life span in a stage of life that almost no other animal experiences! Many researchers see a matrix of evolutionary cause-and-effect conjoining natural selection and cultural evolution in the very existence of your grandma.

# Growing creaky

Life expectancy beyond the reproductive years is dependent to a great extent on genes.

The developments of *senescence* (growing old) are gradual and diffuse. Body parts don't work quite as well as they used to, and they just keep getting worse. For some people, this is a brief stage of life after a long, healthy middle age. Others aren't so lucky.

In particular, the immune system doesn't work as well as it used to, and malformed cells that would have been immediately eliminated at age 35 now may evade immune surveillance and become cancerous.

Age-related changes in the large arteries, as well as cumulative damage to the smaller vessels, can bring about problems with blood pressure.

Inactivity and chronic overconsumption of calories have their worst effects now, in all the major systems.

Still, the brain continues to develop. Research has been shown repeatedly that, under the right conditions, the brain continues to produce new cells and make new connections among neurons in adults as old as 100.

Table 15-1 lists some of the common age-related changes to the body's systems.

**TABLE 15-1**    **Age-Related Changes to the Body's Systems and Associated Health Implications**

| Body System | Change | Implications |
|---|---|---|
| **Cardiovascular system** (see Chapter 9) | Heart increases in size. | There is an increased risk of thrombosis (clotting) and heart attack. |
| | Fat is deposited in and around the heart muscle. | Varicose veins develop. |
| | Heart valves thicken and stiffen. | There is a rise in blood pressure. |
| | Resting and maximum heart rates decrease. | |
| | Pumping capacity declines. | |
| | Arteries decrease in diameter and lose elasticity. | |
| **Digestive system** (see Chapter 11) | Teeth may be lost. | There is an increased risk of hiatal hernia, heartburn, peptic ulcers, constipation, hemorrhoids, and gallstones. |
| | Peristalsis slows. | Rates of colon cancer and pancreatic cancer increase in the elderly. |
| | Pouches form in the intestines (in a condition known as *diverticulosis*). | |
| | Liver requires more time to metabolize alcohol and drugs. | |
| **Endocrine system** (see Chapter 8) | Glands shrink with age, decreasing hormone release. | Numerous homeostatic mechanisms are disrupted. |
| | | The metabolic rate decreases. |

*(continued)*

**TABLE 15-1** *(continued)*

| Body System | Change | Implications |
|---|---|---|
| **Lymphatic system** (see Chapter 13) | Thymus gland shrinks with age. | Cancer risk increases. |
| | Number and effectiveness of T lymphocytes decrease with age. | Infections are more common in elderly. |
| | | Autoimmune diseases (such as arthritis) increase. |
| **Integumentary system** (see Chapter 4) | Epidermal cells are replaced less frequently. | The skin loosens and wrinkles. |
| | Adipose tissue in face and hands decreases. | Sensitivity to cold increases. |
| | There is a loss and degeneration of fibers in dermis (collagen and elastin). | The body is less able to adjust to increased temperature. |
| | Fewer blood vessels and sweat glands are present. | Hair grays and skin becomes paler. |
| | Melanocytes decrease. | Hair thins. |
| | Number of hair follicles decreases. | |
| **Muscular system** (see Chapter 6) | Muscle tissue deteriorates and is replaced by connective tissue or fat. | The muscles lose strength. |
| | Fewer mitochondria are in muscle cells. | Endurance decreases due to fewer mitochondria. |
| | Neuromuscular junction degenerates. | There is a decrease in response and overall function. |
| **Nervous system** (see Chapter 7) | Brain cells die and are not replaced. | Learning, memory, and reasoning decrease. |
| | Cerebral cortex of the brain shrinks. | Reflexes slow. |
| | There is decreased production of neurotransmitters. | Alzheimer's disease occurs in elderly people. |
| | | There is a loss of sensory input (smell, vision, hearing, and so on). |
| **Reproductive system** (see Chapter 14) | *Females:* Menopause occurs between 45 and 55 years of age and causes cessation of ovarian and uterine cycles, so eggs are no longer released, and hormones such as estrogen and progesterone are no longer produced. | Osteoporosis and wrinkling of skin occur, and there is an increased risk of heart attack. |

| Body System | Change | Implications |
|---|---|---|
| | *Males:* Possible decline in testosterone level after age 50; enlarged prostate gland; decreased sperm production. | Impotence and decreased sex drive occur. |
| **Respiratory system** (see Chapter 10) | Breathing capacity declines. | There is decreased efficiency of gas exchange. |
| | Thickened capillaries, loss of elasticity in muscles of rib cage. | Risk of infections such as pneumonia increases. |
| **Skeletal system** (see Chapter 5) | Cartilage calcifies, becoming hard and brittle. | Bones become thinner and weaker. |
| | Bone resorption occurs faster than creation of new bone (loss of bone matrix). | More time is required for bones to heal if they break. |
| | | Osteoporosis risk increases. |
| **Urinary system** (see Chapter 12) | Kidney size and function decrease. | Wastes build up in the blood. |
| | There is decreased bladder capacity. | Incontinence occurs. |
| | The prostate gland in men is enlarged. | The risk of kidney stones increases. |
| | | The urge to urinate is more frequent. |
| | | Urinary tract infections are more likely. |

# 6 The Part of Tens

**IN THIS PART . . .**

Review physics concepts that form the foundation for anatomy and physiology.

Review chemistry concepts important to physiology.

Discover especially interesting facts about the human body.

Chapter **16**

# Ten (Or So) Chemistry Concepts Related to Anatomy and Physiology

B iology is a very special application of the laws of chemistry and physics. Biology follows and never violates the laws of the physical sciences, but this fact can sometimes be obscured in the complexity and other special characteristics of biological chemistry and physics.

This chapter contains a review of some of the principles of chemistry and physics that have special application in anatomy and physiology. Some of these principles overlap — for example, probability is one factor that drives the process of diffusion. Although what follows are oversimplified explanations of very profound and complex matters, we hope they help you better understand anatomy and physiology.

# Energy Can Neither Be Created Nor Destroyed

The *first law of thermodynamics* is that energy can be neither created nor destroyed — it can only change form. Throughout any process, the total energy in the system remains the same. This law is one of the fundamental concepts in physics, chemistry, and biology.

*Energy* is the ability to bring about change or to do work. It exists in many forms, such as heat, light, chemical energy, and electrical energy. Light energy can be captured in chemical bonds, such as in the process of photosynthesis. In physiological processes, the energy in the bonds of ATP is transformed into work when the chemical bonds are broken — to move things, for example, and to generate heat. (And where did the energy in ATP come from in the first place? Ultimately, the sun via photosynthesis.)

Although the total energy in a system always remains the same, the energy available for biological processes does not. Cells can use energy only in certain specific forms. A physiological process that uses ATP doesn't use all the energy stored in those chemical bonds, but the leftover energy isn't in a form that can be used in another physiological process. It is "lost" to physiology, mostly as heat flowing out into the surrounding environment.

# Everything Falls Apart

In our universe, energy is required to create "order" — for example, to build the atomic and molecular aggregations we call "matter" or "stuff." Without continuous input of energy (maintenance), stuff falls apart. No news here for dwellers in the real world. As a physicist might put it, all systems tend toward increasing *entropy* (disorder). This is the *second law of thermodynamics.*

Energy always moves from a point of higher concentration to a point of lower concentration, never the reverse. For example, where two adjacent objects are of different temperatures, heat flows only from the warmer object (higher energy) to the cooler object (lower energy). A state of order contains more energy than a state of disorder because of the energy that went into building the state of order. Energy flows outward into the relative chaos of disorder.

**REMEMBER**

Because living systems are highly ordered, the implications of the second law of thermodynamics are profound for physiology. The law means that *physiological homeostasis* (the maintenance of order) is an active process that requires energy. The energy that must be applied to drive any physiological process comes from releasing the chemical bonds in ATP. It means that physiological reactions proceed in only one direction — they aren't reversible (unlike, say, sodium and chlorine ions that go into solution in water and then reorganize back into salt crystals spontaneously when the water is removed).

The ultimate physiological implication of the second law is the inevitability of death.

# Everything's in Motion

Particles in a solution fly around constantly and collide with one another all the time. This kind of motion is called *Brownian motion.* The higher the temperature, the more frequent and harder the collisions. It's the reason why any reaction that *can* happen *will* happen, because (most of) the particles required for the reaction will collide sooner or later. (But see the "Probability Rules" section that follows.) This is especially important when considering all the molecules (such as glucose and ions) that move through membranes by simple or facilitated diffusion.

Brownian motion is also a mechanism of entropy. Each of the molecular collisions converts energy in the molecules to heat, in which form the energy is transferred to the surroundings.

# Probability Rules

Everything that can happen will happen — *some of the time.* Other times, it won't. The proportion of times it *does* happen depends on a lot of factors. If a solution contains large numbers of each of two molecules required for a reaction, the different types will collide frequently. So, concentration affects the chances that a reaction will actually occur. The higher the solution's temperature, the more frequently molecules will collide and facilitate the reaction. But almost never will every possible reaction actually happen. Just by chance, some of these molecules won't meet up with their counterpart molecule. That's life. The chance, or randomness, can be quantified as *probability.* As with this hypothetical reaction, so with everything else related to biology and physiology: Probability, not certainty, rules.

By the way, the existence of life itself is highly improbable. And the probability of the existence of the uniqueness that is you is more improbable still.

# Polarity Charges Life

A molecule is said to be *polar* when the positive and negative electrical charges are separated between one side of the molecule and the other because of unequal electron sharing. For example, a molecule of water is polar because the oxygen hogs the electrons concentrating the negative charge on the oxygen atom. So the water molecule has a positive charge at one end and a negative charge at the other, similar to a magnet. It attracts and holds other polar molecules. Methane is *nonpolar* because the carbon shares the electrons with the four hydrogen atoms uniformly.

Polarity underlies a number of physical properties of a substance, including surface tension, solubility, and melting and boiling points. In physiology, polarity strongly determines which molecules form bonds and which don't — like how oil and water don't mix. More specifically for the study of physiology, lipids and water don't mix. Living cells use this principle to control the flow of substances into and out of the cell.

*Lipids* are a large and varied group of organic compounds, including fats and oils. All lipids have *hydrophobic* portions to them — that is, they don't mix with water. Why not? Because a lipid is nonpolar, so it can't form bonds with water. Water molecules push nonpolar molecules aside to get closer to other polar molecules.

Visualize a party where some people gather around the TV to watch a game and others congregate in the kitchen. The game watchers are the polar entities (supporters of one team or the other), and the other folks are the nonpolar entities (sharing an interest in nonpolar subjects like biological development and the effect of heat on complex organic substances). To belabor the analogy: The polar entities, after they've taken their positions close to other polar entities (on the couch), tend to maintain their state and position relative to those they've bonded with, while vibrating in place. The nonpolar entities move around relative to one another (milling about the kitchen), and they hold and release each other (children) easily and often. A different set of polar entities (the teenagers) conduct different physiological processes in seclusion from both the nonpolar and the other polar entities.

# Water Is Special

Water is arguably the most important molecule in physiology. It accounts for around 60 percent of an adult's body weight. Water's strong polarity gives it characteristics that make it uniquely suited to providing its numerous functions.

Water has a high specific heat. A substance's *specific heat* is the amount of heat required to raise the temperature of 1 gram of the substance 1 degree Celsius. Because water has a high specific heat, it can absorb heat from our active physiological process without increasing the body temperature.

The polarity of water also separates molecules from each other; dissolving them. This makes it useful as a method of transport (like in blood). This also makes it an ideal environment for chemical reactions to occur. As such, nearly all our metabolic reactions take place in water.

# Fluids and Solids

Physiological processes, generally speaking, take place in fluids, and the properties of fluids are very important in these processes.

In everyday conversation, "fluid" means "liquid," something that's usually water-based, like juice, broth, or tea. In physics and chemistry, though, a watery solution is one kind of fluid, whether it's one you'd care to drink or not. Air is another kind of fluid. Fats are fluids, even when they're solid: Butter is exactly the same substance whether cold or warm, and so is every other form of fat. Technically speaking, glass and pure metals are fluids!

Salt, in contrast, is a solid. Salt (NaCl) crystals flow out of their containers in every kitchen and dining room, yes, but that doesn't make salt a fluid. It's got to do with the molecular structure. In solids, atoms are tightly packed together in a geometrically precise formation called a *crystalline lattice.* Sodium chloride is the model for this: Equal numbers of sodium and chlorine ions, each linked to six other ions, all pull each other in as tightly as the forces of polarity (electrical charge) require and allow. Solids are rigid at the molecular level; once bound together in a crystalline lattice, every atom in the molecule remains in place relative to its surrounding molecules.

In fluids, things move around more. Components come together in various ways — carbon dioxide and molecular oxygen ($O_2$) dissolve from air into water and back into air (in the lungs). Fluids take the shape of their container. Air flows into and fills your alveoli. A watery mass in your stomach changes shape with every churning contraction. Gaseous fluids can be easily compressed because the molecules are already so far apart. However, the compressibility of liquids is very limited because the water molecules are already held together just about as tightly as they can be made to go.

# Under Pressure

Boyle's law describes the inverse relationship between the volume and pressure of a gas. If nothing else changes, such as temperature, an increase in volume brings about a decrease in pressure. When the pressure drops in a fixed space, it creates a vacuum.

The mechanisms of breathing utilize Boyle's law. When the diaphragm contracts, it increases the volume of the lungs, which decreases the pressure. The vacuum pulls air in through the upper respiratory tract. It's also a driving force for the cardiac cycle — opening and closing valves to move blood through the chambers of the heart.

# Redox Reactions Transfer Electrons

The concept of *reduction-oxidation* (or *redox*) reactions is basically this: An electron is transferred from one chemical entity (atom or molecule) to another. The entity that receives the electron is said to be *REDuced.* The entity that releases the electron is said to be *OXidized.* In a redox reaction, the reduction of one entity is always balanced by the oxidation of another. The entities are called a *redox pair.* The redox reaction changes the *oxidation state* of both entities. In some cases, the oxidized entity undergoes another reaction to acquire another electron. Note that this isn't a simple reversal of a redox reaction but a new reaction that involves another electron "donor" and frequently requires an enzyme catalyst.

**TIP**

Here's a clever mnemonic to help with the terminology: OIL RIG — Oxidation Is Losing, Reduction Is Gaining (electrons, that is).

In biological systems, redox reactions are highly controlled and very important. Chemical energy is stored in electron bonds and released (made available for work) by redox reactions. Redox reactions are commonly part of *signaling pathways.* A change in the oxidative state of some molecules carries information. A change in the oxidation state of an entity can affect its polarity, which, in turn, affects its solubility in water and thus its ability to enter or leave a cell through the cell membrane. An entity that becomes more soluble can also become more available metabolically, which can be very important for some metal ions like iron and calcium.

**REMEMBER**

Redox reactions play a crucial role in both of the most important reactions in biology: photosynthesis and cellular respiration. *Photosynthesis* is the reduction of carbohydrate to glucose and the oxidation of water molecules to molecular oxygen, using light energy. (Molecular oxygen is $O_2$ — the oxygen atoms from two water molecules joined together.) In *cellular respiration*, glucose is oxidized to $CO_2$ and $O_2$ is reduced to water.

# Chapter **17**

# Ten Phabulous Physiology Phacts

This merest smattering of the everyday miracles of the anatomy and physiology of this one species inspires awe at the power of evolution's forces. Numerous things make us stand out from our mammalian and primate relatives. Be they clearly evolutionary or of unexplained origin, here are a few for your enjoyment.

## Unique to You: Hands, Fingers, Thumbs

There they are, down at the end of your arms, one on each side, a matching pair, unique to you in their details. Other humans have them, too, but no other animal has them. Your hands are certainly far removed from the front paws typical of most mammals, and they're greatly specialized compared even to those of other primates, including man's closest evolutionary relatives.

One specialization is the *opposable thumb*, which is a thumb that can touch each finger on the same hand. (Go ahead — try it now!) Along with that, the human thumb is *prehensile*, meaning capable of grasping. This anatomy underlies the development of manual dexterity and fine motor skills in humans. The prehensile, opposable thumb makes possible tool making, hunting and gathering, textile and metal crafts, art, writing, cooking, and possibly the very existence of human culture.

# Nothing's Better than Mother's Milk

All the copious research that has been done on the topic over many years has pointed in the same direction: The best nutrition for a human baby is human milk. Human milk is a complex mixture of over 200 different components, and no other substance produced in another animal or yet in a laboratory matches its ability to meet the needs of a human infant. A baby doesn't necessarily need its *own* mother's milk, though. The composition of milk is remarkably consistent — the age, health status, diet, or geographic location of the mother notwithstanding.

Like all foods, the core components of milk are carbohydrates, proteins, and fats. The proportions of these components in the watery matrix and the specific carbohydrate, protein, and fat molecules in a species' milk are precisely adapted to that species' infants' needs. Human milk is the ideal nourishment for a slow-growing, warm-blooded newborn: low in protein (rat's milk has 12 times as much), high in lactose, a sugar with twice the energy content of glucose, and high in the essential fatty acids required for neural development. (As we discuss in Chapter 15, the configuration of the human female pelvis drives the birth of a baby with a relatively undeveloped brain.)

Human milk contains many other substances that affect nutrition and development in different ways. Milk and its precursor, *colostrum*, essentially lend the baby part of the mother's immune system until it can make its own: B cells, T cells, neutrophils, macrophages, and antibodies (see Chapter 13). Lactoferrin and iron-binding protein cause the scant iron in milk to be fully absorbed across the digestive membrane. Milk also has human hormones and growth factors that are believed by some to be required to optimize the development of the brain and other organs.

# It's Apparent: Your Hair Is Different

Along with milk, hair is a defining characteristic of class Mammalia. This class has found hair to be an adaptable accessory and put it to many uses: mechanical protection, UV protection, thermoregulation, sexual selection, social signaling, and waterproofing, among others.

The genus *Homo* is distinguished by an apparent lack of hair. Evolutionary theorists suggest that the early forebears of *Homo* were about as hairy as gorillas, who put their hair to use in all the aforementioned ways, making those *arrector pili* (the tiny muscles that give you goose bumps) all the more useful. What could have driven so drastic a change in so useful an accessory?

Anatomists note that humans haven't "lost" their hair — the skin is covered with hair follicles at about the same density as other apes. But the hair itself is different. Most of it is short and fine, and in some cases barely visible. Head hair is longer and coarser than body hair. Head hair and body hair may be curly. (No other primate has curly hair anywhere.) It may be lightly pigmented or apigmented. How does *Homo* escape a predator or thermoregulate under that? Maybe he runs away on his long, hairless legs, cooled by a steady stream of water from newly evolved glands on his hairless chest and arms. Evaporative cooling would be quite the benefit for a hunter's life on the hot, dry, equatorial savannah.

A typical mammal has dense hair covering the epidermis. As far as thermoregulation goes, wearing a warm blanket is good for retaining heat but bad for dissipating it. Many mammals, including many large predators, rely on panting and certain patterns of behavior to thermoregulate. For example, on hot days, some mammals lie in the shade near the watering hole. The hunter who could be active when others were stressed escaped predation and ate well — not to mention didn't have to worry as much about lice or ticks.

But what about the cold nights? An agile thinker could put his prey's skin and fur to its rightful purpose, keeping a mammalian body warm and safe from abrasion and mechanical shock.

# The Only Thing You Have to Fear Is . . .

The *amygdalae* are paired structures in the middle brain almost exactly the size and shape of almonds, whence they get their name. They've been attracting attention in neuropsychiatry for 60 years — which is pretty much the entire history of neuropsychiatry.

Early research found that neural circuits through the amygdalae connected the midbrain, among the most primitive brain structures, to the frontal cortex, the most advanced. These circuits are part of the limbic system and are thought to be critical in regulating emotion and in guiding emotion-related behaviors.

The amygdalae have been associated clinically with a range of mental and emotional conditions, including depression, autism, and even "normalcy." Physicians have widely and publicly discussed one case in particular — a woman whose amygdalae are partly nonfunctional. This patient is incapable of experiencing the emotion of fear. The doctors have tried everything, not just for research purposes but because a total lack of fear is a maladaptive trait; it threatens her well-being and survival. This patient has been injured and victimized in situations that normal, healthy fear would have kept her far away from.

Some neuropsychiatric researchers theorize that the amygdalae evolved as part of a protective mechanism. Probably very early on in the evolution of vertebrates, the amygdalae reacted to change in the chemical environment, moving the organism away from toxic substances. New organisms adapted this functionality to perceive and respond to new stimuli in the environment. Stepping back from a chemical spill is still a model of appropriate, fear-based reflexive behavior. The circuit between the frontal lobes and the amygdalae is what generates such survival-enhancing behavior.

# You Smell Well!

It is often said that, compared with other animals, the human nose is poor at gathering the information available from volatile molecules in the environment. Just how does the human olfactory sense compare with that of other animals? Take a look at the evidence.

As with other mammals, the human olfactory structures are located at the interface of the brain and the airway. Specialized neurons called *olfactory neurons*, actually protuberances from the brain, sit right on the border of the nasal passages, behind and slightly above the nostrils.

An olfactory neuron bears olfactory receptors on its plasma membrane. An olfactory receptor recognizes a certain chemical feature of an odor molecule, but that feature is present in numerous kinds of odor molecules. The receptor can bind any odor molecules that have that feature. Thus, humans don't have a single receptor for "coffee" or "lavender" or "wet dog." They have many receptors for many kinds of molecules released into the air and drawn into the nose. The brain assembles its olfactory perception of the environment by aggregating the signals from the various receptors. The process is similar to that of vision. Odor recognition is like object recognition based on aggregating many different impulses from the retina. The combination of receptors communicating with the brain gives us our recognition of around 50,000 different scents.

Molecular biology experiments in the early 1990s led to the identification and cloning for research purposes of a large family of olfactory receptors. The experiments showed that the family of genes coding these receptors is the largest in the mammalian genome. Some animals have more than 1,000 different receptors; humans have about 450. One out of every 50 human genes is for an odor receptor!

Based on information from the complex set of olfactory receptors, the brain can determine the concentration of an environmental odor and distinguish a new odor signal from the background odor noise, which is how you get "used to" a smell.

The 450 olfactory receptors give humans their very sophisticated palates, allowing them to enjoy the possibly hundreds of different flavors of a savory meal. Apart from entertainment, this faculty may have helped humans discover new food sources as they moved into new climates and environments.

# Microbes: We Are Their World

For thousands of millions of tiny creatures, your gut is the only universe they know. They live and die in that warm, moist, nutrient-rich, immune-protected environment. They work almost every minute of their lives, providing a service to their community and their universe and abiding by the laws of thermodynamics. These good citizens of the gut are adapted specifically to that environment, in the way of symbiotic organisms, and can survive nowhere else.

The internal tissues — blood, bone, muscle, and the others — are normally free of microbes. But the surface tissues — the skin, the digestive and respiratory tracts, and the female urogenital tract — have distinctive colonies of symbiotic microorganisms. The term *symbiosis* (adjectival form, *symbiotic*) describes a more or less cooperative and reciprocal relationship between organisms and species. Symbioses are, by definition, good for all. That brings us back to your supporting role as Mr. or Ms. Universe.

The microbe colonies derive from the "host" (a particular human body) a steady supply of nutrients, a stable environment, and protection. The host gets help with some tricky digestive tasks, stimulation of the development and activity of the immune system, and protection against colonization by other, less-well-adapted (pathogenic) microbes. Truly, you'd have difficulty digesting and gaining nutrients from the typical human diet without them.

Considering only cell number, you are more bacteria than human, with bacterial cells outnumbering your own by at least ten times. More than 200 species of bacteria are commonly found in one or another of the human symbiotic colonies. The number and species in a microbe colony is influenced by various characteristics of the host, including age, sex, diet, and genetic makeup. So, a question to ponder: If these microbes are, literally, vital to your survival and you are, just as literally, vital to their survival, does that change your understanding of "you" and "them"?

# The Pesky Appendix

Your poor little appendix gets quite the bad rap — labeled with words like *pointless* or *vestigial* (the proper term for an organ with no function). Although the appendix is much smaller than it was in our hunter-gatherer ancestors (in modern humans, it's about 4 inches, located where the small intestine meets the large intestine), calling it useless just isn't quite fair.

Long ago, our diet was full of foliage — lots of leafy greens, nuts, berries, even bark. We don't make, nor have we ever made, an enzyme to break down cellulose, which is the main carbohydrate in plant structures. Because of this, when the *chyme* (digested food) reached the large intestine, it was bulkier, requiring a larger *cecum* (pouchlike first segment of the large intestine). The appendix, also much larger at the time, housed bacteria that made *cellulase,* the enzyme that breaks down cellulose. They got a meal and we got a stool that was easier to pass.

Over time, we started growing our own food and cooking it. We stopped eating so many cellulose-rich foods and started using heat to help break things down. (Compare eating a raw carrot to a cooked one.) The cecum shrunk but maintained its function because it's still the first part of the large intestine. The appendix apparently lost its purpose, as evidenced by its tiny size and lack of any secretions. Hence, its classification as a vestigial structure.

Seemingly supporting this is the lack of complications from appendix removal. Because the appendix is a dead-end tube, it's easy for bacteria to get trapped in there. Especially if it's not one of our normal gut flora, it starts replicating and you have an immune response: inflammation (known as *appendicitis*). The inflammation may resolve on its own, but its isolation increases the likelihood of rupture, allowing your soon-to-be-considered feces into your abdominal cavity. Not good. That's why the most common treatment for appendicitis is removal of the organ.

Recent research has shown, though, that the appendix is composed of lymphatic tissue, indicating an immune function. Further investigation has shown that the appendix acts very much like a "safe-house" for good bacteria. Any time you take an antibiotic or suffer from lower-intestinal distress, you lose some of that beneficial bacteria. If there are always some just hanging out in your appendix, they can easily repopulate your large intestine before a dangerous one can take hold. So, although you can clearly survive just fine without it, your appendix does, in fact, have a function.

# Talkin' about Breath Control

You don't have to think about breathing. The steady in-and-out continues while you sleep and go about your daily business. The depth and rhythm adjust to your level of effort. Just climb those stairs; breathing will take care of itself. Many of us would have died young if breathing required constant attention.

Humans are also capable of controlling their breath. Cetaceans (whales and dolphins) can, too; must, in fact, and some also use breath control to sing. Other animals can't — or at least don't appear to. Canines in chorus are not really controlling their breath.

Humans use breath control to generate speech. Humans can make finely controlled exhalation pass over the vocal cords while the length and thickness of the cords is changing to generate different frequencies of sound. The lips, tongue, glottis, and other structures shape the vibration, allowing you to make those finely distinguished sound symbols called "words" and "syllables." Singing, closely related to talking, requires even finer breath control. We doubt that such a hypersocial, hypercommunicative species as Homo sapiens could have evolved as far as it has without speech and singing.

Various religious practices and breathing disciplines take breath control in another direction. Conventional physiological thinking is skeptical of the idea that the conscious brain can reach "down" and exert control over the quintessentially autonomic processes of breathing. However, brain imaging studies and other experimental results have shown that some people with a long history of regular meditation practice have important differences in their neural systems. Some people believe that systematic breath control exercises can provide benefits to many organ systems: the cardiovascular, digestive, neurological, and endocrine systems for a start.

# Taking Your First Breath

You spent your fetal development with fluid in your lungs. Not a problem because your mother "breathed" for you (not that you had access to air anyway). The fluid is squeezed out during delivery and coughed out shortly after. Some of it is absorbed by the lung tissue itself. So, instantly, with that first scream at birth, the infant's lungs fill with air and begin gas exchange. Why, then, are respiratory issues the main concern for babies born prematurely?

The problem is water. Alveoli that are filled with fluid can't perform gas exchange, but they do require moisture. The nasal cavity warms (to get the oxygen molecules moving faster) and moistens the air as we bring it in. That's why your nostrils feel drier in the winter, the air is drier and pulls more moisture from the lining. The moist air contributes to the thin layer of water that lines the inner wall of the alveoli, allowing for the oxygen and carbon dioxide to move through the walls (between the cells) and into the capillaries. Unfortunately, water has a high surface tension; the individual molecules are highly attracted to each other. Because alveoli are spherical and lined with water, this creates an inward pull from all directions, collapsing them.

A collapsed alveolus obviously won't bring in air for gas exchange. In order to counteract the water's surface tension, thereby keeping the alveoli nice and spherical, the cells secrete *surfactant.* The production of surfactant is one of the final steps in fetal development. In fact, one of the protein components of surfactant is thought to be a trigger for the onset of labor. Babies born prematurely don't yet make surfactant, so they can't breathe on their own. Though the presence of surfactant was discovered in the 1950s, it wasn't until the 1990s that researchers found an effective way to administer it. Since then, the number of premature babies who die from respiratory distress has been cut in half.

# Is Blood Really Blue?

Nearly every diagram of blood vessels printed in color shows the arteries in red and the veins in blue. People with fair skin can look down at their wrists and see very clear, blue veins. So surely that means the blood in the arteries is red and the blood in the veins is blue, right? Not so much.

First, your blood vessels aren't see-through. They have several layers of tissues. The fact that your veins appear blue and your arteries are red (though they're too deep to be seen through the surface of the skin) is related to the color of the tissue layers as well as the color of the blood inside. And although venous blood is a different color from arterial blood, it's definitely not blue.

*Hemoglobin* is the predominant protein in the red blood cells (RBCs). Specialized for the transport of the blood gases, hemoglobin has the capacity to transport molecular oxygen (attached to the heme group) and carbon dioxide (attached to the globin portion) simultaneously.

When it's carrying oxygen molecules, it's called *oxyhemoglobin* and is bright red. When it is not full of oxygen, it's called *deoxyhemoglobin* and is dark red. Between the darker shade of red and the structures of the walls of the veins, you see a blue color through the skin.

# Index

# E

ear wax (cerumen), 84

eccentric contraction, 124

eclampsia, 321

ectopic pregnancy, 321

edema, 250

effectors, 37

ejaculation, 313

elastic cartilage, 92

elastic fibers, 67

electrocardiograms (ECG; EKG), 205–206

electrolytes, 176

embolus, 209, 211

embryo, 62, 328

emphysema, 229

endochondral ossification, 95–96

endocrine glands, 169, 172

endocrine system
  age-related changes to, 337
  glands, 172–183
  hormones, 168–172
  overview, 167
  pathophysiology, 183–186
  roots and word fragments, 12

endocytosis, 50

endometriosis, 319–320

endometrium, 304

endomysium, 123

endoplasmic reticulum, 47, 53

endorphins, 160

endothelial cells, 155

enteric nervous system, 150

enterogastric reflex, 239

enzymes, 56

eosinophils, 278, 285

ependymal cells, 147

epidermal dendritic (Langerhans) cells, 272

epidermis, 38, 75–79

epididymis, 311

epimysium, 123–124

epinephrine (adrenaline), 160, 170, 177–178, 208

epiphysis, 93–94

epithelial cells, 38

epithelial tissue (epithelium), 23, 68–70

erectile dysfunction (ED), 320

erythropoietin, 173, 254

esophagus, 235–236

estradiol, 178

estrogen, 170, 177–178

eukaryotes, 46

eukaryotic cells, 46–54

eupnea, 221

excretion, 254–255

exocrine glands, 172

exocytosis, 50, 171

expiration, 221–222

# F

facial bones, 97–99

facilitated diffusion, 49

$FADH_2$, 31–33

fascicles, 124

feedback, 37

fetal distress, 321

fetus, 328

fibrillations, 210

fibrin, 39

fibrinogen, 39, 191

fibroblasts, 86

fibrocartilage, 92

fibromyalgia, 140

fibrous connective tissue (FCT), 92–93

filtration, 49–50

fingerprints, 80

first law of thermodynamics, 344

fissures, 152

flat bones, 95

flavin adenine dinucleotide (FAD), 31–32

fluid balance, 74, 254, 264–265

fluid-mosaic model of cell membrane, 48–49

fluids, 347

follicles, 304

follicle-stimulating hormone (FSH), 173, 307–308

foramen magnum, 100, 153

foreskin (prepuce), 312

friction ridges, 80

fusiform, 125

# G

GABA (gamma-aminobutyric acid), 160
gallbladder, 243, 251
gallstones, 251
ganglia, 147
gas exchange, 216
gastric (duodenal) ulcers, 247–248
gastric juice, 237, 245, 272
gastric pits, 245
gastrins, 182, 245
gastrocnemius, 140
genes and genetic material
　gene expression, 51, 58, 60
　gene structure, 59–60
　genome, 51, 58, 324
　protein synthesis, 60–61
　traits, 59
germ cells, 298
gingiva, 234, 246
girdles, 103–104
glands
　defined, 172
　enteric endocrine, 180–182
　gonads, 178–179
　hypothalamus, 173–174
　metabolic regulation, 175–178
　pituitary, 173–174
　in skin, 82–84
glial cells (neuroglia), 146–147
glomerular capsule (Bowman's capsule), 256–257
glomerular filtration, 261–262
glomeruli, 256, 261
glucagon, 170, 181–182
glucocorticoids, 170
glucose, 30, 32–35, 45
glutamate, 160
glycolysis, 30–32
glycoprotein hormones, 169
Golgi apparatus, 47
Golgi body, 54
gonadocorticoids, 170, 176–177
gonadotropin-releasing hormone (GnRH), 308
gout, 115
Graves' disease, 185

gross anatomy, 10
growth, 37, 325
growth hormone (GH), 170, 174
guanine (G), 57–58
gums, 234, 246
gyri, 152

# H

hair, 81–82, 350–351
hair follicles, 73, 81–82
haploid cells, 62, 298
Haversian (central) canal, 91
healing process, 39, 86
hearing, sense of, 162–163
heart, 197–201
heartbeat, 205–206
hematopoiesis, 94, 192
hematopoietic stem cells, 278
hematuria, 260
hemocytoblasts, 192
hemoglobin, 191, 356
hemoglobinopathies, 212–213
hemorragia, 211
hemostasis, 209
heparin, 285
hepatitis, 250
herpes, 292–293
heterotrophs, 28
hiccups, 140
hilum, 255, 276–277
hippuric acid, 260
histamine, 281, 286
histologic anatomy, 10
HIV, 292, 319
homeostasis, 27–28, 33–37, 85, 119, 168, 209, 264–266
homeotherms, 34
hormone replacement therapy (HRT), 179–180
hormones, 168–172
human chorionic gonadotropin (hCG), 314
human development
　dimensions of, 324–326
　over lifetime, 326–337
　overview, 323

multipotent (adult) stem cells, 62

muscle spasms, 139–140

muscle spindles, 119, 123

muscle tissue, 23, 70

muscular dystrophies (MD), 139

muscular system

    age-related changes to, 338

    functions of, 118–120

    names of muscles by region, 129–138

    overview, 117–118

    pathophysiology, 138–140

    roots and word fragments, 12

    sliding filament model, 126–128

    types of tissue, 121–125

myelinated neurons, 158

myocardial infarction (heart attack), 210–211

myofibrils, 124

myosin, 126–128

myotonic muscular dystrophy, 139

myxedema, 185

# N

Na$^+$/K$^+$ pumps, 156

NADH, 31–33

nails, 82–83, 88

nasopharynx, 217

negative feedback mechanisms, 37

nephron, 256, 262–263

nerves, 147

nervous system

    age-related changes to, 338

    brain, 150–155

    central nervous system, 148–149

    components of, 144–147

    functions of, 144

    impulse transmission, 156–161

    overview, 143–144

    pathophysiology, 166

    peripheral nervous system, 149–150

    roots and word fragments, 12

    senses, 161–165

nervous tissue, 23, 70

neurons, 38–39, 123, 127, 145, 156–158, 170

neurotransmitters, 158–161

neutrophils, 278, 280, 285

nicotinamide adenine dinucleotide (NAD+), 31–33

NK cells, 278

nociceptors, 161

norepinephrine (noradrenaline), 160, 170, 177–178

nose, 216–217

nostrils, 216

notochord, 9

nuclease, 244

nucleic acids, 57–58

nucleotides, 57–60

nucleus, 46–47, 51

# O

obstructive jaundice, 251

occult bleeding, 212

oligodendrocytes, 147

one gene, one protein model, 59

opposable thumbs, 136, 349–350

orchitis, 318

organ systems, 22–24

organelles, 46–47

organism, 21–22, 24, 27

organs, 22–23

osmosis, 49

osmotic pressure, 49

osseous tissue, 90–92

ossification, 95

osteoclasts, 96

osteocytes, 91

osteons, 91

osteoporosis, 113–114

ova and oogenesis, 40, 44, 298–301, 326–327

ovarian cycle, 307–308

ovaries, 304, 315

oxaloacetic acid (OAA), 31–32

oxidative phosphorylation (electron transport chain [ETC]), 30–33

oxidization, 32

oxytocin, 171, 174, 306

# P

palate, 217

palatine bones, 113

pancreas, 34–35, 169, 180–182, 243–244, 251

tuberculosis (TB), 228–229

tubular reabsorption and tubular secretion, 262–263

tumors, 63

Turner syndrome, 303

## U

ulcerative colitis, 249

ultraviolet (UV) radiation, 77, 79–80, 84, 87

uninucleated cells, 125

uracil (U), 57–58

urea, 259

ureters, 255, 257

urethra, 257–259, 312

urethral blockage, 268

urethral sphincters, 258, 264

uric acid, 259

urinalysis, 260

urinary system

  age-related changes to, 339

  functions of, 253–254

  homeostasis, 264–266

  pathophysiology, 266–269

  roots and word fragments, 12

  structures of, 254–259

  urine, 259–264

urinary tract infection (UTI), 268

urinary tract obstruction, 268

urine, 254, 259–264

urobilinogen, 259

uterine cycle, 308–309

uterine tubes, 304

uterus (womb), 304, 314–315

## V

vacuoles, 47

vagina, 305

valves, 198–199

vas deferens, 311

vasoconstriction, 193

vasodilation, 193

veins, 196–197

ventilation, 215–216

ventricles, 151, 154–155, 197–198, 203–205

vermis, 152

vernix caseosa, 84

vertebrae, 99–101

vestibular glands, 313

villi, 238

vitamin D, 84

vitiligo, 290

vulva, 305

## W

water, 346–347

white blood cells (WBCs; leukocytes), 192, 218, 277–280

wisdom teeth, 234

## Z

zygotes, 44, 62–63, 324, 326–327

## About the Authors

**Erin Odya:** Erin is an anatomy and physiology teacher at Carmel High School in Carmel, Indiana, which is widely regarded as one of Indiana's top schools. She takes pride in her role as an educator.

**Maggie Norris:** Maggie is a freelance science writer living in the San Francisco Bay Area. As Fine Print Publication Services, LLC, Maggie offers contract medical and technical writing services to clients in the pharmaceutical, biotech, and medical technology industries; patient care institutions; and research institutions.

## Dedication

To all my students — past, present, and future — who keep me on my toes and inspire me to keep learning; and to Gabe, who, with his six-year-old silliness and curiosity, distracted me nearly every time I sat down to write — I love you for infinity.

—Erin Odya

To Susan.

—Maggie Norris

## Author's Acknowledgments

Thanks to my family and friends, whom I nagged for advice and whose support in this endeavor is much appreciated.

—Erin Odya

## Publisher's Acknowledgments

**Executive Editor:** Lindsay Lefevere
**Project Editor:** Elizabeth Kuball
**Copy Editor:** Elizabeth Kuball
**Technical Editor:** Kristen Bruzzini

**Production Editor:** Vasanth Koilraj
**Illustrator:** Kathryn Born, MA
**Cover Photos:** © Alexilusmedical/Shutterstock